Recurrent Neural Networks

Fathi M. Salem

Recurrent Neural Networks

From Simple to Gated Architectures

 Springer

Fathi M. Salem
Department of Electrical and Computer
Engineering
Michigan State University
East Lansing, MI, USA

ISBN 978-3-030-89931-8   ISBN 978-3-030-89929-5   (eBook)
https://doi.org/10.1007/978-3-030-89929-5

This Springer imprint is published by the registered company Springer Nature Switzerland AG.
The registered company address is: Gewerbestrasse 11, 6330 Cham, Switzerland

To my dear parents and family, I dedicate this work.

Preface

Deep Learning and Neural Networks have been impacting numerous disciplines and technologies in recent years. They are loosely inspired from the connectivities of the (human) brain in their architectural forms and apply optimization approaches.

In general, they distinguish their approaches in the following:

(1) They form multi-layer and/or multi-stage discrete (recurrent) nonlinear systems: This relates to the need of universal, non-handcrafted, parameterized system that, for instance, in the classification case, partitions the input space using parameterized "highly nonlinear" surfaces or **manifolds**. Note that linear systems or single-layer (or single-stage) systems can only provide hyperplanes to separate regions in the input space.

(2) The optimization results in **non-convex optimization** form: This is due to the (nonlinear) multi-layer or multi-stage nature of the (universal) parameterized architectures (models).

(3) The non-convex optimization formulation is *intentionally* searching for a sufficiently "good" local optimum (minimum), and not (the) global optimum!

   The data-driven approach is summarized as follows. The network uses a performance (or loss) function on the available training data (while evaluating the same function on the statistically similar evaluation data). The evaluation data is a proxy for the (unseen) testing data that are assumed to be statistically similar. The network seeks to obtain the highest performance on the evaluation data and uses that mark as a **stopping criterion**. Beyond the stopping criterion, the network experiences *overtraining*, where the training performance would further increase while the evaluation performance would no longer increase (in fact, it may decrease).

(4) The deep learning approach has been to prioritize innovations in the network architecture (i.e., parameterized nonlinear dynamic mapping) while using the simplest (local) search form, i.e., the stochastic gradient descent and its variants. Techniques that increase the equivalent "step" size (or learning-rate) initially pursue random search and deviate from gradient search to avoid "bad" local minima with corresponding poor performance and then to decrease the learning

rate to approximate gradient search evoking concepts in optimization analogous to increasing and decreasing the "temperature" in, e.g., physics.

Recurrent neural networks and systems are growing as a subfield that is inspired from the prevalent presence of recurrency in the brain. Over the years, since the 1940s, it has borrowed tools from statistics, mathematics, physics, optimization, and numerous disciplines, including computer science and engineering.

While recurrent neural networks are concerned with capturing dynamic (nonlinear) mapping of input sequences to output sequences, they are a form of *finite-time* horizon mappings that unfold to staged (layered) feedforward networks with *shared parameters* across the stages (layers). They are used in their capacity as mappings. They are thus distinctly different from their cousins, known as *feedback neural networks* (including the Hopfield networks), used for storing and retrieving associative memories. *Feedback neural networks* are recurrent neural networks for *infinite-time* horizon and (usually) without (external) input.

New architectural forms of deep learning and neural networks are blending, or even fusing, spatial feedforward networks (including convolutional networks) with temporal recurrent networks. It appears that this fusing will increase in the coming years to enable richer dynamic nonlinear mappings. There is also the possible blending of recurrent neural networks with the so-called feedback neural networks of the associative memory types, i.e., storing raw patterns into the networks as (stable) equilibria or other limit sets. Thus, a deeper understanding of the rigor of recurrent neural networks and their learning frameworks is necessary to charter the next phase of deep learning into neural **artificial intelligence** (AI) systems.

I have taught various courses on neural networks over the years with evolving notes. This book is a byproduct of a course I have developed and taught over the last 6 years at Michigan State University and also online. As was the course, this book aims at balancing the mathematical rigor with the practical necessity and using the empowering open-source computational frameworks (i.e., TensorFlow, Keras, and PyTorch) offered by the Big Tech. I also want to take this opportunity to acknowledge the funding support from the **National Science Foundation** (NSF) for the research and education projects that have enabled the development of parts of the material in this book.

I am indebted to my former students and my numerous colleagues for the inspiring research, scholarship, and teaching that we have been engaged in. I am grateful for learning from all my teachers and pioneers as I remain optimistic about the exciting times of **neural AI systems** that are still unfolding. Finally, and most importantly, I am immensely grateful and indebted to my family and friends and thankful for their continuing support.

Okemos, MI, USA
August 2021

Fathi M. Salem

Contents

Part I   Basic Elements of Neural Networks

1   Network Architectures 3
   1.1   The Elements of Neural Models....................................... 3
      1.1.1   The Elements of a Biological Neuron 3
      1.1.2   Models of a Neuron—Simple Forms 6
      1.1.3   The Connection Parameters: Weights......................... 7
      1.1.4   Activation Functions: Types of Nonlinearities 8
   1.2   Network Architectures: Feedforward, Recurrent, and Deep
      Networks 10
      1.2.1   Feedback Neural Networks: Vector Form 11
      1.2.2   An Alternate View Diagram of Feedforward
           Network Architectures... 13
      1.2.3   Convolution Connection.. 15
      1.2.4   Deep Learning or Deep Networks 18

2   Learning Processes.. 21
   2.1   Adaptive Learning 21
      2.1.1   Learning Types.. 22
   2.2   Optimization of Loss Functions 23
      2.2.1   The Stochastic Gradient Descent (SGD) 24
   2.3   Popular Loss Functions 27
   2.4   Appendix 2.1: Gradient System Basics 31
   2.5   Appendix 2.2: The LMS Algorithm................................... 35

Part II   Recurrent Neural Networks (RNN)

3   Recurrent Neural Networks (RNN).. 43
   3.1   Simple Recurrent Neural Networks (sRNN).......................... 43
      3.1.1   The (Stochastic) Gradient Descent for sRNN:
           The Backpropagation Through Time (BPTT)) 45

3.2   Basic Recurrent Neural Networks (bRNN)—the General Case   48
3.3   Basic Recurrent Neural Networks (bRNN)—a Special Case   54
3.4   Basic Recurrent Neural Networks (bRNN): Summary Equations   57
3.5   Concluding Remarks..   61
3.6   Appendix 3.1: Global Stability of bRNN   61
3.7   Appendix 3.2: Update Laws Derivation Details   64

Part III   Gated Recurrent Neural Networks: The LSTM RNN

4   Gated RNN: The Long Short-Term Memory (LSTM) RNN   71
4.1   Introduction and Background..   71
4.2   The Standard LSTM RNN..   71
      4.2.1   The Long Short-Term Memory (LSTM) RNN   72
4.3   Slim LSTMs: Reductions within the Gate(s)   74
      4.3.1   The Rationale in Developing the Slim LSTMs...............   74
      4.3.2   Variant 1: The LSTM_1 RNN...................................   76
      4.3.3   Variant 2: The LSTM_2 RNN...................................   76
      4.3.4   Variant 3: The LSTM_3 RNN...................................   77
      4.3.5   Variant 4: The LSTM_4 RNN...................................   77
      4.3.6   Variant 5: The LSTM_5 RNN...................................   77
4.4   Comparative Experiments of LSTM RNN Variants   78
      4.4.1   Experiments on the MNIST Dataset   78
      4.4.2   Experiments on the IMDB Dataset   79
4.5   Concluding Remarks..   82

**Part IV   Gated Recurrent Neural Networks: The GRU and The
            MGU RNN**

5   Gated RNN: The Gated Recurrent Unit (GRU) RNN   85
5.1   Introduction and Background..   85
5.2   The *Standard* GRU RNN   86
      5.2.1   The Long Short-Term Memory (LSTM) RNN   87
      5.2.2   The Gated Recurrent Unit (GRU) RNN   88
      5.2.3   The Gated Recurrent Unit (GRU) RNN vs. the
              LSTM RNN   89
5.3   Slim GRU RNN: Reductions within the Gate(s)   90
      5.3.1   Variant 1: The GRU_1 RNN   91
      5.3.2   Variant 2: The GRU_2 RNN   91
      5.3.3   Variant 3: The GRU_3 RNN   91
      5.3.4   Variant 4: The GRU_4 RNN   92
      5.3.5   Variant 5: The GRU_5 RNN   92
5.4   Sample Comparative Performance Evaluation   92
      5.4.1   Application to MNIST Dataset (Pixel-Wise).................   93
      5.4.2   Application to MNIST Dataset (Row-Wise)   96
      5.4.3   Application to the IMDB Dataset (Text Sequence)   98
5.5   Concluding Remarks..   98

6 Gated RNN: The Minimal Gated Unit (MGU) RNN.................... 101
  6.1 Introduction and Background.. 101
      6.1.1 Simple RNN Architectures 102
      6.1.2 LSTM RNN 103
      6.1.3 GRU RNN.. 104
      6.1.4 Gated Recurrent Unit (GRU) RNNs vs. the LSTM RNNs ... 104
  6.2 The Standard MGU RNN.. 105
  6.3 Slim MGU RNN: Reductions within the Gate(s)...................... 106
      6.3.1 Variant 1: MGU_1 RNN.. 106
      6.3.2 Variant 2: MGU_2 RNN.. 106
      6.3.3 Variant 3: MGU_3 RNN.. 107
      6.3.4 Variant 4: MGU_4 RNN.. 107
      6.3.5 Variant 5: MGU_5 RNN.. 107
  6.4 Sample Comparative Performance Evaluation 108
      6.4.1 Sample Comparative MGU RNN Performance 108
      6.4.2 The Network Architecture...................................... 108
      6.4.3 Comparative Performance on the MNIST Dataset 109
      6.4.4 Reuters Newswire Topics Dataset 112
      6.4.5 Summary Discussion 113
  6.5 Concluding Remarks.. 113

References... 115
Index.. 119

Notation

We list the notation used throughout this book. The notation is mainly adopted from the (online) textbook by Goodfellow et al. (2016b) in the spirit of maintaining a unified standard of notation in the deep learning and neural networks community. We thank the authors of Goodfellow et al. (2016b) for initiating and encouraging the adoption of standard notations for deep learning.

Numbers and Arrays

a	A scalar (integer or real)
$\boldsymbol{a}$	A vector
A	A matrix
$\mathbf{A}$	A tensor
I_n	Identity matrix with n rows and n columns
I	Identity matrix with dimensionality implied by context
$e^{(i)}$	Standard basis vector $[0, \ldots, 0, 1, 0, \ldots, 0]$ with a 1 at position i
$\mathrm{diag}(\boldsymbol{a})$	A square, diagonal matrix with diagonal entries given by $\boldsymbol{a}$
a	A scalar random variable
$\mathbf{a}$	A vector-valued random variable
$\mathbf{A}$	A matrix-valued random variable

Indexing

a_i	Element i of vector $\boldsymbol{a}$, with indexing starting at 1
a_{-i}	All elements of vector $\boldsymbol{a}$ except for element i
$A_{i,j}$	Element i, j of matrix A
$A_{i,:}$	Row i of matrix A

$A_{:,i}$ Column i of matrix A
$A_{i,j,k}$ Element (i, j, k) of a 3-D tensor $\mathbf{A}$
$\mathbf{A}_{:,:,i}$ 2-D slice of a 3-D tensor
a_i Element i of the random vector $\mathbf{a}$

Linear Algebra Operations

$A^\top$ Transpose of matrix A
A^+ Moore–Penrose pseudoinverse of A
$A \odot B$ Element-wise (Hadamard) product of A and B
$\det(A)$ Determinant of A

Sets and Graphs

$\mathbb{A}$ A set
$\mathbb{R}$ The set of real numbers
$\{0, 1\}$ The set containing 0 and 1
$\{0, 1, \ldots, n\}$ The set of all integers between 0 and n
$[a, b]$ The real interval including a and b
$(a, b]$ The real interval excluding a but including b
$\mathbb{A} \backslash \mathbb{B}$ Set subtraction, i.e., the set containing the elements of $\mathbb{A}$ that are not in $\mathbb{B}$
$\mathcal{G}$ A graph
$Pa_{\mathcal{G}}(\mathrm{x}_i)$ The parents of x_i in $\mathcal{G}$

Calculus

$\dfrac{dy}{dx}$ Derivative of y with respect to x

$\dfrac{\partial y}{\partial x}$ Partial derivative of y with respect to x

$\nabla_x y$ Gradient of y with respect to $\boldsymbol{x}$

$\nabla_X y$ Matrix derivatives of y with respect to X

$\nabla_{\mathbf{X}} y$ Tensor containing derivatives of y with respect to $\mathbf{X}$

$\dfrac{\partial f}{\partial \boldsymbol{x}}$ Jacobian matrix $\boldsymbol{J} \in \mathbb{R}^{m \times n}$ of $f : \mathbb{R}^n \to \mathbb{R}^m$

$\nabla_x^2 f(\boldsymbol{x})$ or $\boldsymbol{H}(f)(\boldsymbol{x})$ The Hessian matrix of f at input point $\boldsymbol{x}$

$$\int f(x)dx \qquad \text{Definite integral over the entire domain of } x$$

$$\int_{\mathbb{S}} f(x)dx \qquad \text{Definite integral with respect to } x \text{ over the set } \mathbb{S}$$

Probability and Information Theory

$a\perp b$	The random variables a and b are independent
$a\perp b \mid c$	The random variables are conditionally independent given c
$P(a)$	A probability distribution over a discrete variable
$p(a)$	A probability distribution over a continuous variable, or over a variable whose type has not been specified
$a \sim P$	Random variable a has distribution P
$\mathbb{E}_{x\sim P}[f(x)]$ or $\mathbb{E}f(x)$	Expectation of $f(x)$ with respect to $P(x)$
$\text{Var}(f(x))$	Variance of $f(x)$ under $P(x)$
$\text{Cov}(f(x), g(x))$	Covariance of $f(x)$ and $g(x)$ under $P(x)$
$H(x)$	Shannon entropy of the random variable x
$D_{\text{KL}}(P\|Q)$	Kullback–Leibler divergence of P and Q
$\mathcal{N}(x; \mu, \Sigma)$	Gaussian distribution over x with mean μ and covariance Σ

Functions

$f : \mathbb{A} \to \mathbb{B}$	The function f with domain $\mathbb{A}$ and range $\mathbb{B}$
$f \circ g$	Composition of the functions f and g
$f(x; \theta)$	A function of x parametrized by θ. (Sometimes we write $f(x)$ and omit the argument θ to lighten notation.)
$\log x$	Natural logarithm of x
$\sigma(x)$	Logistic sigmoid, $\dfrac{1}{1 + \exp(-x)}$
$\zeta(x)$	Softplus, $\log(1 + \exp(x))$
$\|x\|_1$	L_1 norm of x
$\|x\|_2$	L_2 norm of x (One may use $\|x\|$ instead.)
$\|x\|_p$	L_p norm of x, $1 \le p \le \infty$
x^+	Positive part of x, i.e., $\max(0, x)$
$\mathbf{1}_{\text{condition}}$	It is 1 if the condition is true, 0 otherwise

Sometimes we use a function f whose argument is a scalar but apply it to a vector, matrix, or tensor: $f(\boldsymbol{x})$, $f(X)$, or $f(\mathbf{X})$. This denotes the application of f to the array element-wise. For example, if $\mathbf{C} = \sigma(\mathbf{X})$, then $C_{i,j,k} = \sigma(X_{i,j,k})$ for all valid values of i, j, and k.

Datasets and Distributions

p_{data}	The data generating distribution
$\hat{p}_{\text{data}}$	The empirical distribution defined by the training set
$\mathbb{X}$	A set of training examples
$\boldsymbol{x}^{(i)}$	The i-th example (input) from a dataset
$y^{(i)}$ or $\boldsymbol{y}^{(i)}$	The target associated with $\boldsymbol{x}^{(i)}$ for supervised learning
X	The $m \times n$ matrix with input example $\boldsymbol{x}^{(i)}$ in row $X_{i,:}$

Introduction to the Book

Deep Learning and Neural Networks, which earnestly began in the 1940s, are accelerating advances in **Artificial Intelligence** (AI) techniques. Indeed, the recent convergence of domain applications: (i) relatively large curated training data, (ii) the advent of high-speed and parallel computing—especially, GPUs and now Cloud Computing—and (iii) continuing, and more practical, innovations have propelled deep learning and neural networks into the forefront of technologies and real-world applications.

While most innovations and real-world applications have focused on employing feedforward neural networks (mostly, parameterized staged feedforward mappings), many applications require (time-)sequence processing. Thus, the case for more efficient dynamic parameterized mappings is made, more naturally expressed by *recurrent* neural networks. On the outset, we point out that these recurrent networks work to realize dynamic mappings of a *finite* sequence of inputs to a finite sequence of outputs. They are distinctly different from their cousins, known also as recurrent, or *feedback*, neural networks for associative memory. Whereas recurrent networks are used as mappings from sequence to sequence over a *finite-time* horizon, feedback neural networks for associative memory are over *infinite-time* horizon, typically without external input, driven by their initial states to converge to their (stable) limit sets.

Recurrent neural networks are more technically challenging than feedforward networks, and thus there is a tendency by practitioners to stretch the applicability of feedforward neural networks even beyond their natural fitness. This is motivated from the ease in the technical understanding of feedforward mappings without the complexity and challenges of dynamic recurrency or feedback. For one, in training a forward mapping, using the preferred learning methods of (variants of) the stochastic gradient descent (SGD), the process has no *"dynamic time issue."* In essence, training a feedforward neural network preserves the well-behaved gradient system's properties. That means, analytically at least, the overall system is a *gradient system* that does not exhibit any complex dynamics of oscillations, limit cycles, or chaos! Of course, when one deviates appreciably from a continuous-time model in computational numerical approximations, this guarantee is no longer applicable.

In contrast, recurrent neural networks by their nonlinear and recurrent nature must contend with stability and state boundedness—even over finite-time horizons. The issues of state stability and boundedness, and also of overall system robustness, may arise. Robustness of the dynamic behavior is informally described here as a small random change in the neural network structure and/or the input sequence should not drastically change the generated output sequence or class representation.

This book focuses on recurrent neural networks (RNNs) as a natural extension to feedforward neural networks when recurrency (or feedback) gives rise to state memory and the ability to weave time sequences of data. Time here may be real or fictitious. The point is that the processing now involves sequences; example sequences include sensory measurements, sounds, speeches, sentences or text, language translation, videos, etc. Recurrency brings the concept of dynamic state-memory in the forefront. State-memory in this context denotes capturing internal information about the dynamic processing of impinging input sequences.

When in training mode, the appropriate way is to apply the principled framework of non-convex optimization under dynamic system constraints and apply the Calculus of Variations and the Lagrange Multiplier Techniques. When doing so, the flow of the *generalized* backpropagation through time (BPTT) is generated after setting the boundary conditions appropriately. In addition, this principled framework teaches one how to easily incorporate any additional loss function components, especially on the internal variables and layers, e.g., the activation (or hidden) units and/or the internal state. The systematic presentation in this fashion brings to the fore the simple machinery of (non-convex) optimization with dynamic system constraints and the role of the *lambda* (λ), or the historical *delta* (δ), as the co-state of the dynamic recurrent neural networks. It also teaches the simplicity in applying the same unified techniques to all recurrent neural network families from simple to gated architectures (e.g., LSTM and GRU RNNs).

The book begins from scratch. It should be suitable for a course at the Senior undergraduate or first year graduate level or for a practitioner in deep learning and neural networks. When combined with diverse projects using codes in Python that explicitly include the generalized BPTT, or using any of the Open-Source Deep Learning computational frameworks, it should extend to a one-quarter or one-semester project-based class. The sample Python codes and the example Tensorflow-Keras projects, which the author has employed in his classes, can be obtained from the **book's repository**. The book's repository will also include sample class syllabus, how-to notes, and project descriptions to assist in teaching such a class. The book chapters are arranged as self-contained content that are intended to also be suitable as a reference for the deep learning and neural network practitioners.

The book is divided into four parts. In Part I, we introduce two chapters: Chap. 1 discusses the architectures of neural networks as they can be inspired from the (biological) brain's neuronal networks. The teaching analogy is from the retina's multiple consecutive layers wired through the optic nerve to the visual cortex's five main areas (or groups of layers). One may then justify the various multi-layering architectural connections to an intuitive level. The computational neural layers and architectures, however, are relatively simplistic as they are designed to anticipate scaling of the parameterized cascade of nested systems of equations.

Chapter 2 is dedicated to adaptive and incremental learning, which, for supervised learning, focuses on the stochastic gradient descent, and its graphical realization as the backpropagation (BP) or backpropagation through time (BPTT). The chapter ends with two appendices that highlight simple truths: (i) a crisp definition of what constitutes a *gradient system* using elementary calculus and (ii) the BP algorithm is a direct extension of the historically well-known and popular least-mean-square (LMS) algorithm.

Part II consists of Chap. 3, which is dedicated to the form and mathematics of recurrent neural networks. First, it introduces the so-called *simple* recurrent neural networks (RNNs), followed by the *direct* gradient calculations of the backpropagation through time (BPTT). It identifies potential challenges for the simple RNN (sRNN) that have been referred to in the field as the "vanishing" and "exploding" gradients' phenomena.

This chapter then pivots to remedies that could be incorporated to remove the limits on the sRNN. The basic RNN (bRNN) is introduced, promoted by system redesigns that would separate the state-memory (or state) from the present incoming sequence instance. Moreover, these redesigns—adopted from system theory—are to enhance the capacity of the bRNN to prevent the occurrence of the "vanishing" and "exploding" gradients. This chapter sets the formulation of adaptive learning in the principled form of an adaptive non-convex optimization with network constraints. It applies the principles of Calculus of Variations systematically with clear assumptions to obtain what we call the *generalized* BPTT. This technical formulation is general and is applicable to all forms of recurrent neural networks, including the gated (i.e., LSTM) RNNs in the subsequent chapters. The chapter ends with two technical appendices. One shows that the bRNN is in fact bounded-input-bounded-state (BIBS) stable, meaning that if incoming sequences are bounded, then the internal signals and the generated sequences would also be bounded. The second appendix collects more explicit technical derivations for the *generalized* BPTT using the approach of (non-convex) optimization under dynamic systems' constraints.

The remaining parts (Parts III and IV) focus on gated recurrent neural networks (gated RNNs). Part III has one chapter, Chap. 4, that describes the main workhorse of gated RNN, namely the Long Short-Term Memory (LSTM) RNN. The insertion of gate variables in the input path, the output path, and the recurrent (forget) path indeed brings flexibility in the RNN design. The choice in the design of LSTM RNN was to define the gate variables by replicating the structure of sRNN for each gate, driven by *all* the same available signals–external input, bias, and hidden activation and/or (cell) memory signals. The chapter then highlights the redundancy in this construct. It follows by introducing reductions in the gating internal structure—and parameters—and thus brings forward graded parameter (and hence computational) reductions. While in the open literature there are more reduction realizations, we refer to them as *slim* RNN, the chapter introduces only five reductions in the gating signals.

The next section dedicates itself to sample comparative performance among the *standard* LSTM and the five *slim* LSTM family. The summarized study uses the classical IMDB public dataset to exhibit the relative comparative performance of the standard LSTM RNN vis-a-vis the variant slim LSTMs. While more datasets have been used to demonstrate the comparative effectiveness in the open literature, it suffices here to illustrate their relative comparative performance on the classic IMDB dataset. The book repository will include references to applications using other datasets.

Part IV focuses on the other two derived gated RNNs, which embody architectural downscaling of the independent gate signaling. Chapter 5 focuses on the Gated Recurrent Unit (GRU) RNN. Upon direct comparison with the standard LSTM, the GRU RNN introduces two innovations: (i) it reduces the number of independent gates to two by coupling and (ii) it drops the number of nonlinearities in the signal paths to one. Moreover, it effectively feeds back the (cell) state vector, as opposed to the activation "hidden" vector, into the gating signal paths. (Recall that the activation vector is a nonlinear function of the state vector.) It relates the input and the recurrent (forget-) gate signals in a coupling convex direct sum to 1. It is noted, however, that only (ii) is unique to the GRU RNN. In the literature, comparison to the standard LSTM RNN has shown competitive performance on several dataset case studies. It is accepted that the GRU RNNs are different families of architectures and their behaviors may differ in a specific application. As in Chap. 4, this chapter then introduces reduced forms of the GRU RNN in the gating signal's internal parameters. To identify these variant forms, they are called *slim* GRU RNNs. The chapter ends with comparative studies of the *standard* GRU RNNs and some *slim* GRU RNN variants.

Chapter 6 focuses on the so-called Minimal Gating Unit (MGU) RNN, which downscales the GRU RNN and proposes using a single independent gating signal for all three gates. Namely, it uses the one external gate signal to generate the input, recurrent (or forget), and output gating signals. While its contribution is technically minor, it is worth inclusion as it brings a practical architectural reduction in the gating signals to a single source. However, its performance vis-a-vis the GRU RNN remains under evaluation. Similarly, this chapter describes multiple *slim* MGU RNN(s) that reduce the parameters within the single external gate. Finally, the chapter ends with comparative performance among the *standard* MGU RNN and the *slim* MGU RNN variants on example public datasets.

Part I
Basic Elements of Neural Networks

A Note to the Reader Part I collects basic elements of neural networks and deep learning: architectures, common nonlinear functions used for activations, and popular loss functions. It also includes the mathematical derivatives of such functions that are useful in the derivations of the dominate learning approach of the (stochastic) gradient descent. A fluent expert in neural networks may view the two chapters in this part as preliminary material for the core material on recurrent neural networks starting in Chap. 3.

Chapter 1
Network Architectures

This chapter describes the building blocks of neural network architectures. Our (present) understanding of neurobiology is commonly invoked as a guidance (and justification) in structuring the building blocks of (artificial) neural *architectures* (Kandel et al., 2021). And then, mathematics and engineering intervene—as necessary—to understand or ensure practical (machine implementable) functionality and performance. To that end:

(i) One incorporates what is known (and hypothetically, understood) from neurobiology in terms of functionality and then uses (mathematical) abstraction and/or graphs to construct computational architectures for modules or machines.
(ii) One should always be cognizant of possible variations for potential abstracted architectures that (a) may be supported from a neurobiological viewpoint and/or (b) may become useful in introducing (more capable) future alternate architectures.

1.1 The Elements of Neural Models

1.1.1 The Elements of a Biological Neuron

While there is a diverse population of neuron types (Kandel et al., 2021), the elements of a basic neuron are generally understood to be local causal signal flow through the following elements: (i) cell-body (or soma), (ii) axon, (iii) synapse (or synaptic gap), (iv) dendrites, and then to (v) cell-body of typically other neurons. A simple depiction is shown in the cartoon figure of a "typical" neuron, Fig. 1.1.

In more elaborations (Kandel et al., 2021):

(i) The cell-body (or soma) is said to transform multiple analog potentials emanating from usually multiple dendrites. When the aggregated (analog

© The Author(s), under exclusive license to Springer Nature Switzerland AG 2022
F. M. Salem, *Recurrent Neural Networks*,
https://doi.org/10.1007/978-3-030-89929-5_1

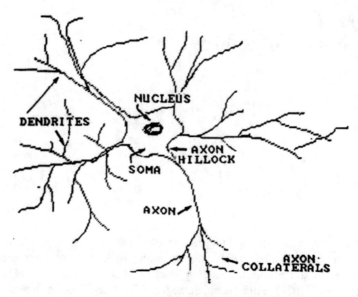

Fig. 1.1 Sketch of simplified typical neurobiological neuron elements

potential) signal exceeds a threshold level, the cell-body initiates a string of (discrete) spikes at the start of its axon—known as the axon hillock. The key point is that the (discrete) spikes are invariant in shape, while their (firing) rate—or more precisely, their total phase (i.e., the integral of the firing rate)—is proportional to, or reflective of, the collective signal into the cell-body. In engineering and mathematical modeling, often the (discrete firing) rate is expressed as an analog real value to represent the output variable of a neuron—even though the physical signal itself is a firing of discrete **invariant** spikes (or discrete pulses).

(ii) The axon initiates the stream of spikes at its hillock and transmits the invariant spiking signals along the axon to its end(s), activating the release of neurotransmitters from vesicles into one or many synaptic gaps (of typically other neurons). The neurotransmitters flowing in the synaptic gaps are of two dominate types, namely, excitatory and inhibitory, to the receiving dendrites' membranes.

(iii) Synapses are believed by neuroscientists to be the localized learning elements (or sites) where changes or plasticity takes place. A synapse comprises the synaptic gap (i.e., separation) between two elements, typically between an axon tip and a dendritic tree membrane. Of course, there are instances where the connections occur between two axons or two dendrites, or other more complex forms.

(iv) A dendrite (or a dendritic tree) receives signals from other neuron unit axons (or from external signal stimulus) by way of synapses (or synaptic gaps), see Fig. 1.2. The signals flow in the synaptic gap(s) is carried through by chemical

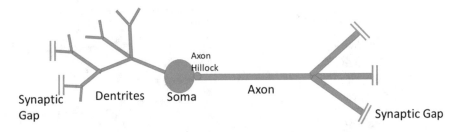

Fig. 1.2 Simplified diagram of the elements of a typical neuron

molecules and is analog (continuous-valued) in nature. These molecules culminate in the creation of analog voltage potential on the receiving dendrite's membrane at a connection location/site. As there may be numerous location sites, these graded electrical analog potentials are *integrated and transformed* along the dendritic tree and collectively impinge on the cell-body.

Examples of well studied neurobiological module networks that justify the current popular engineered architectures include front-end sensory modules of the retina, the olfactory, and the cochlea. Perhaps the retina and the subsequent **visual cortex** are currently an excellent example that encompasses the diverse organizations and processing of neuroprocessing.

The retina, as the front-end visual sensory module, is composed of apparent layering of neuronal units. As in Fig. 1.3, the photocells in the back-end, composed of cones and rods, convert light signals into electrical form, and (feed)forward through the first layer units, to the **bipolar cells layer**, then onto the **ganglion cells layer** to the **optic nerve**.

It should be noted that there are also **lateral connection** layers, identified as the **horizontal cells (layer)** and the **amachrine cells (layer)**. Such lateral connection layers appear to perform local spatial inter-computations among units within one layer—they are usually not *explicitly* emphasized as layers in engineered computational architectures of neural networks and deep learning. However, the spatial inter-computations are approximated by convolutional layers (Bengio et al., 2015; Goodfellow et al., 2016a).

The retina's processed signals feed into the optical nerve(s) to the **visual cortex** (VC) areas and layers. The VC constitutes several layers with relatively distinct functionality. They are simply called visual cortex areas 1–6 (namely, V1–V6). Some, such as V4, may include several subdivisions into identifiable layers and clear pathways of feedback to prior areas (e.g., to V2). In addition to strong feedforward and feedback (recurrent) connections, there are also temporal (i.e., time) functional dependency of the areas' functionality. Moreover, there is the clear spatial and temporal interactions in this VC system. There are numerous hypotheses, supported by neurobiological observations, that early layers (e.g., V1 and V2) perform local spatial and temporal filtering and feature extractions.

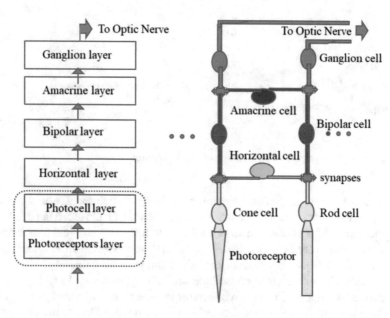

Fig. 1.3 Simplified cartoon diagram of the retina layered networks

The main takeaways from the vast research in neurobiology and related fields (Kandel et al., 2021) are that most of these findings are descriptive in nature (observational), with limited prescriptive experiments. The observations indeed justify the network organization into layers with attributes including the following:

(i) **Local connectivity**—Each of the billions of neurons (for humans, there are estimated to be 10^{11}, or 100 billion neurons), each is connected to the order of 10^4 other neurons

(ii) **Strong feedforward paths** in connectivity among areas and layers

(iii) **Selective recurrent feedback** paths among the areas and layers

(iv) **Lateral interconnections** among units within a layer—this is commonly not explicitly and distinctly present in the engineering mathematical modeling

(v) **Attention processing** that enables refocusing on certain spatial layer regions related to stimulus (e.g., image) signals

(vi) **Temporal dependency** that exhibits time-dependent layer functionality leading to spatio-temporal processing of scenes over time.

1.1.2   Models of a Neuron—Simple Forms

The trend has been to formulate what is understood (or perceived) from neurobiology into mathematical (abstracted) models to capture functionality and behavior.

While neuroscientists are interested in detailed modeling that replicates the neuronal behavior at various degrees of details, engineers and computer scientists are interested in functionality at a scaled network level that replicates the overall behavior of neural **artificial intelligence** (AI) systems. Thus modeling of elements of neurons are often simplified in order to attain a computationally viable model in anticipation of scaling to larger networks. For example, in computational models, dendrites and axons are reduced to just "wiring" connections. Briefly, the simplified modeling trend has resulted in the following:

(1) **Dendrites:** a wire; however, for neurobiology, it involves more complex modeling.
(2) **Soma or cell-body:** a threshold or (smoothing) saturation function. Examples include sigmoid functions such as the logistic function or the hyperbolic tangent function (tanh (.)), or more recently, the **rectified linear unit** (reLu) function. It is noted that the **spiking neuron** function is more physically accurate from neurobiological observation. There is interest in using such function in hardware implementations but is less used in (software) computational modeling.
(3) **Axon:** a wire connection; however, for modeling neurobiology, there are more complex **transmission-line** forms.
(4) **Synapse:** Often abstracted into a single parameter ("weight") per synapse for the usual machine computational models. In neurobiological functionality, however, an engineering **transfer function** or filter would be more accurate. In fact, a dynamic nonlinear ordinary (or even partial) **differential equation** (o.d.e.) modeling would be more suitable to capture the transient and time-spatial flow of neurotransmitters through the synaptic gap. However, that would be at the expense of scaling of the overall network model, and thus non-existent in machine neural AI modeling!

1.1.3 The Connection Parameters: Weights

The connection weights between neurons represent the synapse that may be a simple scalar, a vector, a matrix, or a tensor, Riley et al. (2006) and Boyd and Vandenberghe (2018). A weight modulates with signals in varied ways from multiplications to convolution, to stochastic modulation (Boyd & Vandenberghe, 2018; Riley et al., 2006). Here are some example forms:

1. **Multiplication**
   The connection weights can modulate the signals by performing multiplications from previous neurons to subsequent neurons (or from prior values of a neuron to its present values in the form of feedback). They may be in the form of multiply-and-add operations that result in signal matrix multiplications, or in certain cases in point-wise (Hadamard) multiplication.

2. **Convolution**

The connection may be represented by convolution operations over a fixed sub-field as it is used in convolutional neural networks. The convolution operation allows the network to be applied to non-specified lengths or sizes of signal or images from external sources or prior layers. Whereas a matrix multiplication requires knowledge of the input (or prior) signal size (dimension), convolution operation specifies the sub-field or the **kernel size** but not the input signal/image/tensor size or dimension. This provides flexibility for a designed network to be applied to arbitrary input dimensions.

3. **Stochastic Connection**

There may be connections representing stochastic links with certain probability (Riley et al., 2006) where a neuron is connected to a portion of previous layer neurons with a chosen probability. A analogous approach is used in training of networks via the so-called **dropout layer**. It is noted, however, that this stochastic behavior is removed after training. In the inference (processing) phase when the parameters are frozen at the attained values at the end of training, this stochastic (dropout) layer is removed.

1.1.4 Activation Functions: Types of Nonlinearities

Another important architectural modeling is the choice of the so-called activation function representing the cell-body (i.e., soma) for transforming aggregated signals to a neuron unit output. Historically, there have been the linear and threshold (or hard-limiter) functions that have been replaced by the (differentiable) logistic and the hyperbolic tangent functions. The transition to differentiable functions has been driven by the use of the parameter gradient descent adaptation algorithms that technically require derivatives in order to derive update laws of the parameters (weights and biases). We list the more common activation functions in recent use, and their (closed form) derivatives, in the following. It is noted that these derivatives are needed in the derivations of the gradient descent algorithms. Thus, one can refer to them here when needed in the next chapters. Moreover, the derivatives of these functions are easily expressed in software implementations of deep learning:

1. The linear activation function:

$$x \longrightarrow a(x) = Ax \tag{1.1}$$

where A is any square matrix, including the identity. Of course in the case of the identity, it simply means there is no explicit activation function.

The derivative of the linear activation function:

$$\frac{d}{dx}a(x) = a(x)' = A \tag{1.2}$$

2. The logistic (or sigmoid) function (we use $a(\cdot) = \sigma(\cdot)$):

$$x \longrightarrow \sigma(x) = [1 + \exp(-\gamma x)]^{-1}$$
$$= \exp(\gamma x)/[1 + \exp(\gamma x)] \qquad (1.3)$$

where γ impacts the slope of the logistic function around the origin.
The derivative of the logistic function is

$$\frac{d}{dx}\sigma(x) = \sigma(x)' = (-1)[1 + \exp(-\gamma x)]^{-2}(-\gamma)\exp(-\gamma x)$$

$$= \sigma(x)^2[(\gamma)\exp(-\gamma x)]$$

$$= (\gamma)\sigma(x)^2[1 + \exp(-\gamma x) - 1]$$

$$= (\gamma)\sigma(x)^2[\sigma(x)^{-1} - 1]$$

$$= (\gamma)\sigma(x)[1 - \sigma(x)] \qquad (1.4)$$

The final form of the derivative is given in terms of the function itself. A very appealing property when using the derivatives in software codes!

3. The (vector) softmax function:

$$x = [x_1, \ldots, x_n] \longrightarrow a_i(x_i) = \exp(\gamma_i x_i) \Big/ \left[\sum_{j=1}^{n}\exp(\gamma_j x_j)\right]$$

$$= 1 \Big/ \left[1 + \left(\sum_{j\neq i}\exp(\gamma_j x_j)\right)\exp(-\gamma_i x_i)\right]$$

$$= \left[1 + \left(\sum_{j\neq i}\exp(\gamma_j x_j)\right)\exp(-\gamma_i x_i)\right]^{-1}$$

$$= [1 + (\beta_{-i})\exp(-\gamma_i x_i)]^{-1} \qquad (1.5)$$

where γ_i is a constant and β_{-i} is used to compressedly represent the sums for the exponentials not containing x_i.
The derivative of the Softmax function is
For each i, the derivative is

$$\frac{d}{dx_i}a_i(x_i) = \frac{d}{dx_i}([1 + (\beta_{-i})\exp(-\gamma_i x_i)]^{-1})$$

$$= (-1)[1 + (\beta_{-i})\exp(-\gamma_i x_i)]^{-2}(\beta_{-i}\exp(-\gamma_i x_i)(-\gamma_i))$$

$$= (\gamma_i) a_i (x_i)^2 [(a_i (x_i))^{-1} - 1]$$
$$= (\gamma_i) a_i (x_i) [1 - a_i (x_i)] \tag{1.6}$$

which is similar to the logistic function derivative for each component activation. It is clear that the softmax function is a multi-input version extension of the logistic function.

4. The hyperbolic tangent $tanh(.)$:

$$x \longrightarrow a(x) = tanh(x) \quad = [\exp(\gamma x) - \exp(-\gamma x)]/[\exp(\gamma x) + \exp(-\gamma x)] \tag{1.7}$$

where the hyperbolic function can be decomposed into a difference between a logistic function and a mirrored logistic function.

The derivative of the hyperbolic tangent can be derived in terms of itself as

$$\frac{d}{dx} a(x) = a(x)' = \gamma[1 - a(x)^2] \tag{1.8}$$

5. The rectified linear unit (ReLu)

$$x \longrightarrow a(x) = reLU(x) = max(0, x) \tag{1.9}$$

The derivative of the ReLu is defined as

$$a(x)' = 0 \quad \text{if} \quad x \leq 0$$
$$= 1 \quad \text{if} \quad x > 0 \tag{1.10}$$

1.2   Network Architectures: Feedforward, Recurrent, and Deep Networks

In connecting (abstract models of) single neurons, there is the view that **layering** organizes the process as observed in the neurobiology of the brain—e.g., in the retina and the subsequent areas and layers of the visual cortex. Thus, a group (or a population) of neurons forms a **layer**, which typically interacts with other layers of neurons. While in biology there are also lateral connections within layers, for simplicity, this is often not explicitly used in machine computational models. Therefore, usually one considers feedforward and/or recurrent (i.e., feedback) neural architectures among layers. More recently, the combined feedforward and feedback architectures have been used in (sequential) cascade, usually feedforward, followed by recurrent—or feedback—architectures. Even more recently, new architectures are introduced that merge the feedforward and feedback connectivities into one holistic architectures.

1.2.1 Feedback Neural Networks: Vector Form

The common practice is to use a layer of neurons. This layer typically forms an array of units (more recently, in computational frameworks, one uses a 3D array or a **tensor**). Then units in this layer receive connection from a prior layer and provide connection (outputs) to a subsequent layer. (We can also think vertically as in biology so connections are from a lower array (layer) to a higher array (layer).) Typically, there are no lateral connections in the computational models within an array.

Let us label the single layer notational vector form as follows:

$$x_t = W\bar{s}^p + b \tag{1.11}$$

$$a_t = f(x_t) \tag{1.12}$$

where t is the instance, stage, or layer, $\bar{s}^p$ is the m-d incoming p-th pattern/signal-sample vector at stage t, x_t is the n-d "summing" vector, and a_t is the corresponding n-d vector of the activation function $f(\cdot)$. The matrix parameter W and the vector parameter b will have the necessary compatible dimensions; specifically, W is an $n \times m$ matrix, and b is an $n \times 1$ vector. Note that the first equation is linear and the optional nonlinearity is relegated to the second, activation equation.

This is a memoryless single layer neuron model. That is, the input sample $\bar{s}^p$ impacts the variable x_t in the corresponding layer (or stage) t.

As we anticipate the parameter to be updated at every (instant) stage or layer, we shall (i) index the parameters also, and (ii) to create a cascade of feedforward layers in a sequential form, we shall view the input $\bar{s}^p$ as coming from a previous stage. We can now consider a cascade of N multi-stages (or multi-layers) of such blocks, ordered by the stage (layer) index t, from stage (layer) 1 to layer N. We express the multi-layer feedforward neural networks as

$$a_0 := \bar{s}^p$$

$$x_t = W_t a_{t-1} + b_t \quad t = 1, \cdots, N \tag{1.13}$$

$$a_t = f(x_t) \quad t = 1, \cdots, N \tag{1.14}$$

$$z^p := a_N \tag{1.15}$$

This indeed represents a mapping from the input pattern $\bar{s}^p$ to the network's output z^p. These equations can be equivalently graphically expressed in the cascaded diagram in Fig. 1.4.

Remark 1.1 We use the superscript p to denote the input pattern or sample applied, while we use the subscript index t to denote the layer or the stage of the network. That is, for each p pattern or sample, the network will propagate through its N stages before it outputs a value that corresponds to the p-th input sample. This output is denoted by $z^p := a_N(\cdot)$. Contrary to the prevalent assertions in the deep

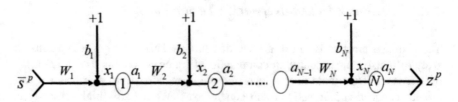

Fig. 1.4 Diagram of N-stage multi-layer feedforward neural network

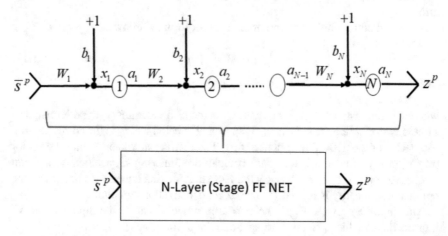

Fig. 1.5 Diagram of N-stage multi-layer feedforward neural network represents a dynamic mapping block

learning field, feedforward neural networks are not static mappings, but they are in fact **finite-horizon** (stage) dynamic mappings.

Remark 1.2 We intentionally used the layer index t as a subscript to denote the layer or the stage. We do so in anticipation of reverting to using the index t as time (real or fictitious) when considering recurrent neural networks. (Please be cognizantly aware that other references or sources may use the layer indexing as a superscript instead).

For illustration purposes the finite-horizon dynamic feedforward network can be illustrated as a (parameterized) mapping from the input sample $\bar{s}^P$ to the output (at the final N-th layer or stage) z^P. This is illustrated in Fig. 1.5.

In Eq. (1.13), we set the initial state condition to the input sample, i.e., $a_0 := \bar{s}^P$. This initial condition then triggers the forward propagation of the "state" dynamics to the end of the finite-horizon stage N to produce the activation state $a_N(\ )$, which is taken to be the network response output z^P.

1.2.2 An Alternate View Diagram of Feedforward Network Architectures

Instead of setting the initial state of the network to be the input pattern, one can follow an alternate setting for feedforward neural networks. We can preserve the external signal to be an input to the first stage or layer, i.e., we set $s_1 = \bar{s}^p$, and set the initial activation state to zero. We shall also have separate paths (or terms) for the activation state and the input, each with its distinct parameter matrix. We shall preserve the (matrix) parameter W_t for the input signal path (or term) and introduce a new (matrix) parameter, namely, U_t, for the activation state path. The changes are applied to generate the following (equivalent) architecture:

$$h_0 := 0$$

$$s_1 = \bar{s}^p, \quad s_t = 0, \quad 2 \le t \le N$$

$$x_t = U_t h_{t-1} + W_t s_t + b_t \quad t = 1, \cdots, N \tag{1.16}$$

$$h_t = f(x_t) \quad t = 1, \cdots, N \tag{1.17}$$

$$z^p := h_N \tag{1.18}$$

where we used the label h_t, instead of the label a_t for the activation function for this distinct architecture. The equations can be represented graphically by the following diagram in Fig. 1.6.

Now a small tweak to the architecture of Eq. (1.16) and Fig. 1.6 helps to convert the finite multi-stage feedforward architecture into a *finite-horizon* recurrent neural network architecture. We shall (i) free up the input to be any sequence to be applied at every stage, (ii) force the dimensions of the weights and biases among all the stages to be the same, and (iii) "share" the weights and biases among all the stages by **folding** the network and view the index t as a (fictitious or real) time index instead

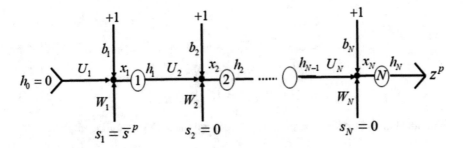

Fig. 1.6 Alternate diagram of the N-stage multi-layer feedforward neural network

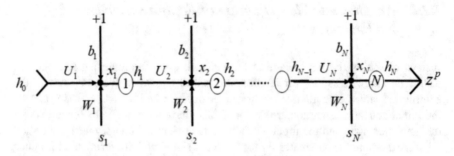

Fig. 1.7 Unfolded simple recurrent neural network architecture

of spatial stage index. These actions will result in the *finite-horizon* recurrent neural architecture as follows:

$$h_0 := \text{I.C.}$$

$$x_t = U_t h_{t-1} + W_t s_t + b_t \quad t = 1, \cdots, N \tag{1.19}$$

$$h_t = f(x_t) \quad t = 1, \cdots, N \tag{1.20}$$

$$z^P := h_N \tag{1.21}$$

which now admits any input sequence s_t applied at every index t. Figure 1.7 depicts the **unfolded** form of a simple recurrent neural networks of N-stage time horizon. This establishes the equivalence of *finite* multi-stage (feedforward) neural networks and *finite-horizon* recurrent neural networks in this particular case.

Finally, we illustrate a typical architecture of a *simple* recurrent neural network followed by a linear (fully connected layer) indexed with time over a finite-time horizon as described by Eqs. (1.22) and (1.25). The corresponding graphical representation of the architecture with its time-unfolding over the finite-time horizon is illustrated in Fig. 1.8

$$h_0 := \text{I.C.}$$

$$x_t = U_t h_{t-1} + W_t s_t + b_t \quad t = 1, \cdots, N \tag{1.22}$$

$$h_t = f(x_t) \quad t = 1, \cdots, N \tag{1.23}$$

$$z_t = V_t h_t + c_t \quad t = 1, \cdots, N \tag{1.24}$$

with an alternate optional output passing through the *softmax* nonlinearity as

$$p_t = \text{softmax}(z_t) \quad t = 1, \cdots, N \tag{1.25}$$

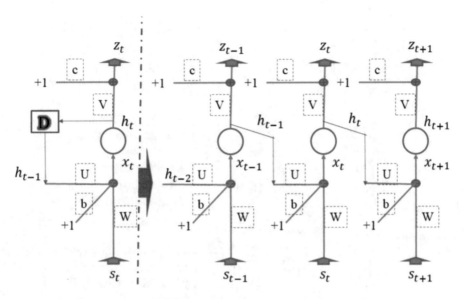

Fig. 1.8 sRNN and unfolding over finite-time horizon

1.2.3   Convolution Connection

The convolutional connection in a layer of a neural network is inspired from the visual cortex in the brain (Fukushima, 1975, 1980; LeCun et al., 1998), but also from filtering in engineering and computer science which has been practiced in numerous areas including filter designs and computer vision processing. Applying convolution operation enables one to reduce the fully connected operation, represented by a matrix or a tensor, which must be compatible to incoming input signal/frame size, to a much smaller **kernel size** (or a **receptive field** region or a sub-matrix). Moreover, this removes the need for knowing, a priori, the full size of the input to the convolution operation. This may be useful when one applies the same architecture to varied-dimension input signal. Convolution operations also reduce the number of parameters (in the kernel) that need to be updated in the learning (i.e., training and evaluation) phase. Convolution can also be applied to 1D, 2D, or 3D signals (emanating, e.g., from speech, image or video data, or live sensors).

1.2.3.1   Convolutional Operations in 1D

We graphically illustrate the convolutional operations applied to 1D example signals (e.g., sensor measurements or speech signals), see Fig. 1.9. This type of convolution is used frequently in **convolutional neural network** (ConvNet) operations to generate successful trained mappings for detection and classification of images

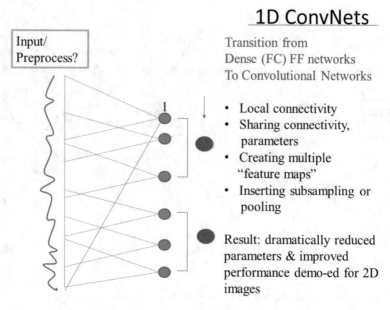

Fig. 1.9 Transition from Fully Connected (FC) Layer to Convolutional Layer

or data. It is, however, easier graphically to illustrate the processing steps on 1D signals.

Assume the 1D signal is the input depicted in analog form on the left of Fig. 1.9, samples of this signal form the 1D input vector. For a fully connected network, each neuron, marked in the figure by the number 1, is connected to every element of the input vector. If the input has 100 dimensions, then this calls for 100 weights. If one reverts to local connectivity, say by 10 parameters, this will use multiple neurons to cover the input elements. For simplicity, assuming no overlapping, this will require 10 neurons in total to cover the input elements (of course, more if there is overlap in coverage). No savings there in parameters. Now, however, we introduce *sharing* of the parameters, i.e., the 10 neurons now use the same 10 parameters. Consequently, we use each input sample *spatially* for training the shared parameters. Thus one now has reduced the parameters to be updated to 10 from 100! Next the pooling operation requires no parameters. One may "pool" together several of these 10 neurons into one, e.g., by averaging their values, or by taking the max. In the figure, 3 neurons are pooled into one, and thus one reduces the number of neurons from 10 to say 3 or 4. The pooling operation reduced the neurons for the next stage (or layer) of operations. This discussion was applied to one neuron from a fully connected layer, to a locally, shared connections. That is, this discussion focused on creating one feature from the input signal.

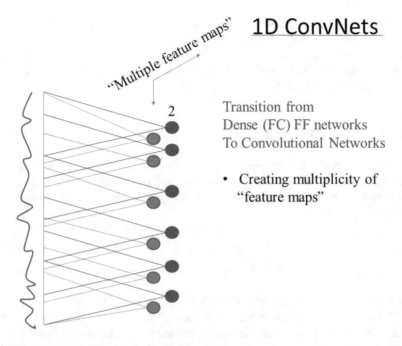

Fig. 1.10 Creating Multiple features

One may repeat these operations for another feature, feature 2, as shown in Fig. 1.10. The new red column of neurons repeats the process to potentially extract another feature with another locally connected and shared *kernel*.

Let us now consider an example recurrent architecture with convolution operations in the following 1D example recurrent network:

$$x_t = [U] * h_{t-1} + [W] * s_t + b \tag{1.26}$$

$$h_t = f(x_t) \tag{1.27}$$

where t, as before, is the (fictitious or real) time, s_t is a vector of dimension m, x_t is the (internal) state of dimension n, and h_t is the activation state vector of dimension n also. Here b is a bias vector of dimension n. Now $[W]$ is a matrix of dimensions $n \times m_1$, where m_1 is $\leq m$. Similarly, $[U]$ is also a sub-matrix of dimension $n \times n_1$, where n_1 is $\leq n$. Typically, the kernel size m_1 is much less than the input vector s_t size m, and the kernel size n_1 is much less than the activation state h_{t-1}. Often the convolution operation may be used in conjunction with regular matrix operations.

Other reduction operations, such as pooling, can also be applied. Simply stated, pooling is a way of transforming several (scalar) outputs of a (typically) convolution layer into one value. In 2D, e.g., a 2×2 convolutional layer output is transformed into one output. Popular transformations used for pooling include (i) averaging (or mean) and (ii) max-value (or ∞ norm).

1.2.3.2   Tensor Processing

One may consider datasets represented by tensors (typically, 3D Arrays). That includes data that is naturally 3D tensors or videos (which are 2D images with the time axis as the third dimension. An example recurrent neural network may be formed as follows:

$$X_t = [U] * H_{t-1} + [W] * S_t + B \qquad (1.28)$$

$$H_t = F(X_t) \qquad (1.29)$$

where the dimensions must be compatible for a meaningful definition.

These architectures have initially been started by cascaded blocks of convolutional neural networks, often optimized for feature representations or classifications, followed by block(s) of recurrent neural networks for sequence-to-sequence mappings. More recently fused architectures have become more popular which integrate the convolutional block(s) within the recurrent neural networks architectures. Advanced architectures can easily be formed on the various open-source deep learning computational frameworks/platforms, e.g., TensorFlow-Keras and PyTorch.

1.2.4   Deep Learning or Deep Networks

Deep learning is essentially deep sequential (or multipath) cascading of diverse layers. The art of building an architecture has been simplified by the computational platforms into cascading of configured blocks, called layers, into compatible connections organized in multitude of ways. The building blocks include convolutional layers, pooling layers, dropout layers, fully connected layers, recurrent layers, bi-directional recurrent layers, which include LSTM layers, GRU layers, etc. A designer can arrange these building block layers into any architectures. There are numerous such (optimized) architectures in the open-source platforms mostly developed from intuitions and trial and error. Figure 1.11 below depicts a cartoon sketch of a cascade of compatible deep (diverse) layers, shown from layer 1 to layer N, representing a full deep network, or even one branch of multi-branch deep network.

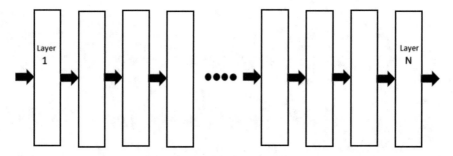

Fig. 1.11 Sketch of a cascade of layers for a full (or a branch of) a deep neural network

Chapter 2
Learning Processes

Once an architecture of a neural network is formed or chosen, there will be a set of (i) *parameters*, i.e., the parameters comprising the matrices and vectors (or tensors) in the network to be determined, often incrementally and adaptively, by a learning (i.e., training and evaluation) process, and (ii) the so-called *hyper-parameters*, i.e., a set of other parameters typically specified by the user or relegated to separate (training and evaluation) strategies (e.g., slower adaptive strategies on different time scales including supervisory controls and/or genetic algorithms).

2.1 Adaptive Learning

Learning in neural networks is associated with the incremental modifications or changes of the parameters. (This simplistically mimics the role of the biological synapses). In its simplest form, learning is represented by the changes in values of the parameters (i.e., the "weights" and "biases") of a neural network. Learning thus is the process of incremental changes to the weights and biases based on acquired sensory signals or (training) data in order to achieve a desired objective. The desired objective is usually captured by a chosen performance measure, often called a **loss function**.

The learning process is usually partitioned into two phases: (i) a training and evaluation phase and (ii) a (testing/deployment) inference phase. These phases may be iterated at select stages over the network's life cycle.

© The Author(s), under exclusive license to Springer Nature Switzerland AG 2022   21
F. M. Salem, *Recurrent Neural Networks*,
https://doi.org/10.1007/978-3-030-89929-5_2

2.1.1 Learning Types

There are several categories of *learning* approaches. We list here the prominent types in brief descriptions.

Supervised Learning

It is the dominant approach when the network (or the user) has access to input(s) and their corresponding desired labels as (training and evaluation) data. The learning algorithm would entail defining a suitable measure (often called a loss function) and would carry out a form of **stochastic gradient descent** (SGD) to incrementally change the parameters in order to achieve pre-specified acceptable performance level of the loss function. It is noted that there are numerous forms of SGDs employed in order to improve the learning (i.e., training and evaluation) performance. It should also be noted that the family of SGDs are the defacto favored learning algorithms for their relative simplicity and suitability to graphical implementations convenient for the leading deep learning computational frameworks.

Typically, there are distinct objectives for supervised learning:

(1) Classification: Input signals/data are classified into categorical (discrete) classes/labels.
(2) Detection: Input signals/data (or images) are used to detect the presence of objects/classes within the images.
(3) Regression: Input signals/data are mapped to corresponding desired output values via a "continuous" input/output map that extends the capability to interpolation and extrapolation. That means the (trained) network becomes a continuous (nonlinear dynamic) map from input to output.

Unsupervised Learning

Technically, learning here should solely rely on the input data/signal without (a priori known) corresponding labels. The role of learning is to extract information from the input data itself in order to classify, detect, or correctly map to corresponding classes, values, or clusters. An example is the adaptive networks that perform **principal component analysis** (PCA) (Haykin, 2009) and/or **independent component analysis** (ICA) (Waheed & Salem, 2003) decomposition of the input data. While these methods purport to be pure unsupervised learning, one however uses measures or loss functions—without direct external labels—to guide the process. Along these approaches fall the (data) clustering methods, and also the so-called auto-encoders that use multi-layer neural networks with conceptual approaches analogous to the PCA techniques.

Hebbian Learning

Unsupervised learning also includes the so-called *Hebbian learning* that has many forms and realizations, loosely motivated from neuronal biological observations (Kandel et al., 2021).

Hebbian learning stems from a postulate of learning observed by D. Hebb in 1949 that is quoted as:

> When an axon of cell A is near enough to excite a cell B and repeatedly or persistently takes part in firing it, some growth process or metabolic changes take place in one or both cells such that A's efficiency, as one of the cells firing B, is increased.

The main point of this observation includes the following attributes of the observed learning process:

- Learning appears to be a local activity among cells.
- It may have the form of coupling, interaction, or correlation.
- It may depend on timing of spiking pulses or synchronization; it is a time-dependent activity.
- There are many more potential interpretations of such a general observation, however.

This general observation of neuronal cell behavior has resulted in numerous mathematical forms and expressions attempting to capture its essence. The results include simple mathematical (correlation) expressions applied to the weight (and bias) updates for "curve fitting" of this general observation.

Reinforcement Learning (RL)

In the real world, one assumes that a system/agent/actor/entity/network interacts with its "environment." One thinks of a creature, a robot, etc., interacting with (or probing) its surroundings and "learning" from its interactions. Reinforcement learning seeks to capture this process of *interacting* with the environment. In this case the network is not passive as it receives signals, it in fact actuates a probing or exploration action to induce a response from the environment, then process the response in order to generate a new probing signal, or improve the previous action signal. In essence RL involves sensing and action or control in sequential interactions. While motivated from numerous sources, RL is best addressed with the framework of **approximate dynamic programming** (ADP) (Bertsekas, 2019; Vrabie et al., 2012). For our purposes, where a network aims at developing a mapping estimator from sensors (or inputs) to classes or labels, RL is outside the scope of this book.

2.2   Optimization of Loss Functions

Once constructed, a neural network constitutes a multi-stage *dynamic* mapping. Then, one employs a chosen *loss* function to guide its optimization—typically minimization.

A loss, or a cost, function is a *non-negative scalar* function used as a measure of performance (e.g., minimum loss error or cross-entropy or alternatively maximizing accuracy). The loss function of a neural network is an implicit function of all its parameters.

Let the loss function be given in general as a (non-negative) scalar function:

$$L(y, p) := L(y_t, p_t(s_t, \theta)) \tag{2.1}$$

where θ is a generalized vector of *all* scalar parameters in the network. Its elements are the elements from all the (parameter) vectors and matrices of the layers that comprise the network. Here, y_t is the label/class vector (or even matrix, or tensor in some applications), and $p(\cdot)$ is the network response or "output" vector corresponding to the input vector s_t. The response vector $p(\cdot)$ often denotes the likelihood estimate or probability of belonging to the label/class of its element index. That is, the element $p_i(\cdot)$ is the probability of belonging to the label/class i.

It is noted that "learning" in a multi-layer neural network forms a *non-convex* optimization with dynamic system constraints and that it would generally have multiple (local) optimal points. It is also noted that in neural networks one is not interested in global optimization but rather a local optimization toward an *acceptable* minimum point of the loss function. That is, minimizing the loss to within a sufficiently small range that would correspond to an acceptable range of performance, say in terms of accuracy. Simplistically, one can view the local minima in two categories: (i) good local minima and (ii) bad local minima. Any good local minimum that satisfies the acceptable (e.g., accuracy) performance is acceptable. Of course, robustness and repeatability of performance, attained from using large training data and procedures, is a plus.

2.2.1   The Stochastic Gradient Descent (SGD)

The stochastic gradient descent (SGD) is by choice the main approach that (the deep learning community) has been using to steer the (incremental) update of the parameters toward an *acceptable* local minimum (of course, assuming it exists). Why that is the choice? One reason is because, in the context of the **Neural Networks and Deep Learning** (NNDL) formulations, SGD and its variations work satisfactorily well. All the fancy innovations and improvements in performance in **NNDL** have been focusing on the developments of *architectures*. Examples of such architecture developments are the deep layering in convolutional neural networks (e.g., the AlexNet, GoogleNet, ResNet, etc.) and the gated recurrent neural networks (e.g., **Long Short-Term Memory** (LSTM) recurrent neural networks).

The gradient descent is the negative of the gradient of the scalar loss function with respect to the parameter vector of interest, namely θ. The gradient defines the vector of maximal change from the present (level curve) of the **loss function**.

We shall recall the **directional derivative** along a direction defined by a unit vector, say $\hat{r}$, in the parameter space (Boyd & Vandenberghe, 2018; Riley et al., 2006). The directional derivative is formally defined as follows:

$$dL_\theta(y, p(s_t, \theta)) = \lim_{\epsilon \to o} [L(y, p(s_t, (\theta + \epsilon\,\hat{r}))) - L(y, p(s_t, (\theta)))]/\epsilon = \nabla L \cdot \hat{r} \tag{2.2}$$

where $\hat{r}$ is a (unit) directional vector along which the derivative is calculated, ϵ is a scalar that goes to zero in the limit, and $\cdot$ here represents a dot product.

Remark 2.1 In practice, this derivative may be approximated by its computational derivative by letting ϵ be sufficiently small, denoted by $\delta\epsilon$ as follows:

$$dL_\theta(y, p(s_t, \theta)) \approx [L(y, p(s_t, (\theta + \delta\epsilon\,\hat{r}))) - L(y, p(s_t, (\theta)))]/\delta\epsilon \qquad (2.3)$$

This numerical approximation to the derivative is often used in practice to verify that the analytical derivative based on Calculus is correct.

For multi-layer feedforward neural networks the whole system may become a *gradient* system. This is the case for all feedforward neural networks—including convolutional neural networks (ConvNets).

For (feedback) recurrent neural networks (RNNs), when the RNN is used as a mapping of a *finite* sequence to a *finite* sequence, it may also be cast as a gradient system if the dynamics of the network can be made decoupled (e.g., by substantially different time scales) from the dynamics of the gradient descent parameter adaptation.

Remark 2.2 If the recurrent neural network is used as a dynamical system, for an infinite-time horizon, then the overall system could not necessarily be a gradient system! It has the potential of being a *complex dynamical system* that may exhibit complex behavioral internal dynamics such as oscillations and chaos!

Definition
The gradient of a scalar function (of the parameters) at a point $\bar{p}$ is a vector in the (hyper-dimensional) parameter space pointing in the direction of maximum *rate of change* and whose magnitude is equal to the maximum *directional* derivative.

Physical Interpretation
Let an equi-potential surface S be defined by the loss $L = $ constant. The surface S is defined by its orthogonal unit vector, say $\hat{s}$ at any point, say, $\bar{p}$. Compute the directional derivative at the point $\bar{p}$ of L. This directional derivative is dependent on a vector to define its direction in the high-dimensional parameter space.

The incremental change from the equi-potential surface is defined as follows:

$$\Delta L(\bar{p}) = \{\nabla L(\bar{p}) \cdot \hat{r}\} \cdot \hat{s}$$

which is equal to the projection of the directional derivative along the (hyper-surface) unit vector $\hat{s}$.

This means that the incremental change or move from the equi-potential surface S is zero if the derivative direction $\hat{r}$ is orthogonal to the unit vector $\hat{s}$, i.e. this will be along the surface S tangent vector. In contrast, the incremental change is maximum if the direction of the derivative $\hat{r}$ is along the **surface direction** $\hat{s}$ itself, i.e., orthogonal to the (tangent of the) equi-potential surface S. The *gradient* is defined to be the maximum directional derivative of the loss $L(\cdot)$.

For the gradient (descent) in the parameter space of the vector θ, one writes

$$-\nabla_{\theta_t} L_t = -\frac{\partial L_t}{\partial \theta_t} \tag{2.4}$$

This gradient (descent) vector may be scaled appropriately to define the change to be applied to the parameter vector θ from instance t to instance $t + 1$ as

$$\Delta \theta_t = -\eta \frac{\partial L_t}{\partial \theta_t}$$

$$\theta_{t+1} = \theta_t + \Delta \theta_t \tag{2.5}$$

where η is the scaling factor that controls the amount of change along the (descending) gradient vector. We note that the scaling factor η can be any sufficiently small scalar quantity that must be non-negative for gradient descent.

The scalar η may be replaced with non-negative quantity that is a function of time. That is, we may have η_t instead to denote that this scaling factor is a function of the time step t. We shall make a factual statement that all the (scalar) modifications to the SGD must always be non-negative (or better yet, be positive) quantity for the gradient descent to retain its characteristics of always descending along trajectories toward a local minimum.

Examples of variable η_t that have been successful in applications and have been incorporated into most open-source computational platforms are:

1. **RMSprop**: which divides a constant value, say η_0, by a (running) moving average of the square root of the (absolute value of) elements of the gradients over a specified time interval (Chollet, a; Tieleman & Hinton, 2012).
2. **Adam**: qualitatively similar to RMSprop but uses scaled running averages of both the gradient and the root square to obtain an estimate of the gradient descent direction and magnitude at the present time t (Chollet, a; Kingma & Ba, 2015).

   Both of these numerically motivation stochastic gradient descent approaches have faired well in practice in numerous experiments and applications of **NNDL**.

   We now add a different variable learning rate that has worked well in some applications but does not have very wide use. It is called the **exponential-loss** learning rate. We describe it in the following (see Ahmad & Salem, 1992; Dey & Salem, 2017a):
3. The **exponential-loss learning rate**: Here one chooses the learning rate to be

$$\eta_t = \eta_0 \, exp(\gamma L(\cdot)) \tag{2.6}$$

where η_0 and γ are two judiciously chosen positive scalar constants, and $L(\cdot)$ is the loss function used in the experiment or application. Note that one may also use here a moving or running average of the loss function.

This third choice of η_t has an *analytical justification* for a gradient and it has an intuitive appeal. In essence, it makes the effective (gradient) stepping large if the loss function is (still) large, and it automatically reduces the learning rate if and when the loss function becomes small. There are no pre-set number of iterations or epochs for learning rate down scaling!

Assume that initially the loss value is large due to the fact that the randomly initialized weights are far from an optimal for the dataset being processed, which is a reasonable assumption. Then, the learning rate would be relatively large resulting in parameter-update stepping away in large strides from the current parameter position. This random walk continues till the loss function decreases in value toward performance-improved parameter values. As the learning rate decreases further, this would render the update law closer to becoming a true (continuous-time) gradient descent whereby it would decrease to the (nearest) local minimum.

2.3   Popular Loss Functions

The loss functions used in **neural networks and deep learning** (NNDL), include all L_p norms, of course. However, L_2, L_1, and L_∞ norms are more frequently used. While the L_2 norm has been used traditionally in supervised learning to seek to minimize the output layer error vis-a-vis the given label, the L_1 norm can be used for hidden units to capture some essence of **sparsity** or **statistical independence** of the internal (hidden) activations (Waheed & Salem, 2003).

The following are the examples of popular loss functions for supervised learning in practice. We present each loss function followed by its mathematical derivative with respect to the variables of interest. The derivative expressions would be helpful reference in the upcoming derivations of the parameter update using the (stochastic) gradient descent (SGD):

1. The **Least Mean Square** or L_2 norm squared loss function

$$L = \frac{1}{2}\sum_{i=1}^{n}(p_i - y_i)^2 = \frac{1}{2}[p - y]^T[p - y] \tag{2.7}$$

where the index i is over all elements of the output vector p and the label y. In most applications, the vector p uses a **softmax** nonlinearity, and its dimension represents the number of classes or categories where each indexed element's value represents a (probability) estimate of belonging to the indexed class.

Its element-wise derivative is

$$\frac{d}{dp_i}(L) = (p_i - y_i) \tag{2.8}$$

$$\frac{d}{dz_i}(L) = \frac{dL}{dp_i}\frac{dp_i}{dz_i} = (p_i - y_i)p_i(1 - p_i) \tag{2.9}$$

$$\frac{d}{dw_{ij}}(L) = \frac{dL}{dp_i}\frac{dp_i}{dz_i}\frac{dz_i}{dw_{ij}} = (p_i - y_i)p_i(1 - p_i)\frac{dz_i}{dw_{ij}} \tag{2.10}$$

where we used the derivative of the *softmax* or the *logistic* functions given by $p_i(1 - p_i)$, see Eq. (1.6). Note that this derivative block approaches zero if p_i approaches zero or 1.

2. The **cross-entropy** (CE) or logloss function (let $L := CE$):

$$L = -\sum_i y_i \, log(p_i) \tag{2.11}$$

where the index i is over all output elements of the vector p. We observe that the CE loss function is related to the following definition of the **Kullback–Leibler divergence** (KLD) and only differs from it by a constant.

$$KLD := E_y(log(y/p)) = \sum_i y_i \, log(y_i/p_i) \tag{2.12}$$

$$= \sum_i y_i \, log(y_i) - \sum_i y_i \, log(p_i) \tag{2.13}$$

Note that the first term is a constant (as a function of the parameters) since it is a function of the labels only. Note also that a constant in a loss function does not change the profile (it only shifts the profile), and its contribution to the derivative is zero. It thus does not change the gradient (i.e., the derivative) of the loss function. Consequently, it does not alter the update law based on the stochastic gradient descent (SGD).

Its derivative is

$$\frac{d}{dp_i}(L) = -y_i(1/p_i) = -(y_i/p_i) \tag{2.14}$$

$$\frac{d}{dz_i}(L) = \frac{dL}{dp_i}\frac{dp_i}{dz_i} = -(y_i/p_i)p_i(1 - p_i) = -y_i(1 - p_i) \tag{2.15}$$

$$\frac{d}{dw_{ij}}(L) = \frac{dL}{dp_i}\frac{dp_i}{dz_i}\frac{dz_i}{dw_{ij}} = -y_i(1 - p_i)\frac{dz_i}{dw_{ij}} \tag{2.16}$$

An interesting observation in this derivative is that there is a cancellation of the p_i term in the definition of the derivative of the softmax (or the logistic function for 1D output). This derivative block is no longer equal to zero at p_i equal to zero.

3. The **binary cross-entropy** (BCE) :

Let $L = BCE$.

$$L = -E\{y \, log(p) + (1 - y) \, log(1 - p)\}$$

$$\approx -\sum_{i=0}^{i=n}\{y_i \, log(p_i) + (1 - y_i) \, log(1 - p_i)\} \tag{2.17}$$

We begin with the derivative with respect to the output p_i. As p_i is (commonly) a softmax, the derivative is computed from the output itself! Thus one gets

$$\frac{d}{dp_i}(L) = -[y_i(1/p_i) + (1 - y_i)(-1/(1 - p_i))]$$

$$= -[(y_i(1 - p_i) - (1 - y_i)p_i)]/p_i(1 - p_i)$$

$$= -[y_i - p_i]/p_i(1 - p_i) \tag{2.18}$$

$$\frac{d}{dz_i}(L) = \left(\frac{dL}{dp_i}\right)\frac{dp_i}{dz_i}$$

$$= -([y_i - p_i]/p_i(1 - p_i))p_i(1 - p_i) = -[y_i - p_i] \tag{2.19}$$

$$\frac{d}{dw_{ij}}(L) = \frac{dL}{dp_i}\frac{dp_i}{dz_i}\frac{dz_i}{dw_{ij}}$$

$$= -[y_i - p_i]\frac{dz_i}{dw_{ij}} \tag{2.20}$$

In this case, one observes that the derivative of the BCE has contributed precisely the inverse expression of the derivative of the softmax and thus cancels it completely.

In Chap. 1, we introduced the commonly used activation nonlinearities, particularly in the (last) output layer. The (derivative of each of these) nonlinearities impact the gradient descent form in mitigating the update of the parameters in the "landscape" of the parameter space. For example, the derivative of the softmax (or in 1D, the logistic) function produces the expression that decays to zero as the softmax approaches 0 or 1, and it maxes out at the midpoint (between 0 and 1), namely, at $1/2$. For ease of review and access, we repeat here the definition of the softmax, the now frequently used nonlinearity in deep learning classifications to mimic the probability estimate of the i-th element of the output belonging to class i.

The Softmax

$$x = [x_1, \ldots, x_n] \longrightarrow a_i(x_i) = \exp(\gamma_i x_i)/[\sum_{j=1}^{n} \exp(\gamma_j x_j)]$$

$$= 1/[1 + (\sum_{j \neq i} \exp(\gamma_j x_j)) \exp(-\gamma_i x_i)]$$

$$= [1 + (\sum_{j \neq i} \exp(\gamma_j x_j)) \exp(-\gamma_i x_i)]^{-1}$$

$$= [1 + (\beta_{-i}) \exp(-\gamma_i x_i)]^{-1} \qquad (2.21)$$

where β_{-i} is used to denote the sums over variables not containing x_i.

The derivative of the softmax function:

For each i, the derivative is

$$\frac{d}{dx_i} a_i(x_i) = \frac{d}{dx_i}([1 + (\beta_{-i}) \exp(-\gamma_i x_i)]^{-1})$$

$$= (-1)[1 + (\beta_{-i}) \exp(-\gamma_i x_i)]^{-2}(\beta_{-i} \exp(-\gamma_i x_i)(-\gamma_i))$$

$$= (\gamma_i) a_i(x_i)^2 [(a_i(x_i))^{-1} - 1]$$

$$= (\gamma_i) a_i(x_i)[1 - a_i(x_i)] \qquad (2.22)$$

which, for each component activation, it is analogous to the (1D) logistic function derivative. This derivative has a max at $a_i = 1/2$ and decays smoothly to zero at $a_i = 0$ or 1.

In this book, we shall use this *simple* recurrent neural network architecture:

$$x_t = U h_{t-1} + W s_t + b, \qquad\qquad t = 0, \ldots, N \qquad (2.23)$$

$$h_t = g_t(x_t), \qquad\qquad t = 0, \ldots, N \qquad (2.24)$$

$$z_t = V h_t + D s_t + c, \qquad\qquad t = 0, \ldots, N \qquad (2.25)$$

$$p_t = \psi_t(z_t), \qquad\qquad t = 0, \ldots, N \qquad (2.26)$$

This simple recurrent neural network (sRNN) sets the notation we are adopting in this book. Here s_t is the external input sequence, x_t is the internal state, h_t is the activation state (which is a nonlinear function of x_t), z_t is the linear layer output, and p_t is the nonlinear output. Note that $\psi(\cdot)$ is typically the softmax function. At time step $t = 0$ this network requires the initial activation state value h_{-1} and the input sequence $s_t = s_0, \cdots, s_n$ in order to generate the internal state x_t (and activation state h_t) and the final output sequences p_t.

All the variables and functions will be introduced in detail in the next chapter.

2.4   Appendix 2.1: Gradient System Basics

We summarize here the key features of what constitutes a gradient system. Consider a generic n-dimensional real variable vector θ. (Mathematically, one expresses it as $\theta \in R^n$.) Consider a non-negative scalar loss function L that is a function (explicitly or implicitly) of this generic variable θ, namely $L(\theta)$. We remark that in the domain of (supervised) neural networks and deep learning the generic variable θ here represents all the network parameters that would need to be updated. The update is commonly executed using variations of the (stochastic) gradient descent (SGD).

A continuous-time *gradient* system is defined as follows:

$$\dot{\theta} = \pm \frac{\partial L}{\partial \theta} \tag{2.27}$$

That is, the continuous-time rate of change of the vector θ is equal to the (positive or negative of) *gradient*, which is the partial derivative of the loss L with respect to the vector θ. We specifically use the term gradient *descent* if the negative sign is chosen, and thus one is seeking a (local) minimum in the "landscape" of the parameter space. Analogously, we use the term gradient *ascent* if the positive sign is chosen, and thus one is seeking a (local) maximum in the "landscape" of the parameter space. One may insert any positive scalar, say γ, to control the speed of the gradient descent or ascent, *tracing* the "landscape" profile toward the nearest optimal point. In the continuous-time version of gradient descent, there is no jumping along the profile or skipping, and the θ trajectories will move while maximally decreasing the loss value and eventually converge to a local optimal point.

Similarly, one can define a gradient-like (descent) system as follows:

$$\dot{\theta} = -\frac{\partial L}{\partial g(\theta)} \tag{2.28}$$

where $g(\cdot)$ is an invertible function of θ, i.e., it has a unique inverse function. Example $g(\cdot)$ functions include the nonlinear logistic (or "sigmoid") and hyperbolic tangent (*tanh*) functions.

Formal Statement
1. Assume the non-negative scalar loss function $L(\cdot)$ is a *radial* and positive definite function and is used to define a gradient descent system.

    If $L(\theta) \geq \beta(\theta)$ $\forall \theta \notin$ bounded region B,
    where $\beta(\theta)$ is a *class-k* function.

    Then all trajectories converge to the bounded region B. (This statement follows from the more general **Liapunov Theory**.)
2. If the scalar function $L(\cdot)$ is simply bounded below (by a finite number).

    Then, all trajectories of its generated gradient system must converge to a stable equilibrium point closest to the initial condition (i.e., a local minimum). If $L(\cdot)$

is not bounded below, then all trajectories either converge to a local (stable) equilibrium point, or go to $-\infty$.

In case θ is a vector, then the (continuous-time) gradient descent update is compactly expressed as

$$\dot{\theta} = -\nabla_\theta L := -\frac{\partial L}{\partial \theta} \tag{2.29}$$

which can be expressed in (component-wise) equations as

$$\dot{\theta}_1 = -\nabla_{\theta_1} L := -\frac{\partial L}{\partial \theta_1} \tag{2.30}$$

$$\dot{\theta}_2 = -\nabla_{\theta_2} L := -\frac{\partial L}{\partial \theta_2} \tag{2.31}$$

$$\vdots$$

$$\dot{\theta}_i = -\nabla_{\theta_i} L := -\frac{\partial L}{\partial \theta_i} \tag{2.32}$$

$$\vdots$$

$$\dot{\theta}_n = -\nabla_{\theta_n} L := -\frac{\partial L}{\partial \theta_n} \tag{2.33}$$

where $\theta \in R^n$; $\theta = (\theta_1, \cdot, \theta_n)^T$.

We note that the time rate of change of the loss function $L(\theta)$ along a trajectory $\theta(t)$ (in the parameter space) becomes

$$\frac{dL}{dt} = \sum_{i=1}^n \left(\frac{\partial L}{\partial \theta_i}\right)\left(\frac{d\theta_i}{dt}\right) \tag{2.34}$$

Using the definition of what constitutes a gradient descent system as expressed by Eq. (2.32), one gets

$$\frac{dL}{dt} = -\sum_{i=1}^n \left(\frac{\partial L}{\partial \theta_i}\right)^2 \leq 0 \tag{2.35}$$

Equation (2.35) is the unique property of gradient (descent) systems. It says that the rate of change of the loss function along any trajectory of the gradient descent system is always decreasing or equals 0. Implicitly, it is decreasing along the gradient (descent) that is orthogonal to the tangent of the equi-potential level surfaces of the loss function. These statements will become clearer with the elaborations below. Note, for a gradient system, trajectories are always decreasing (more precisely, non-increasing) along trajectories for any scalar loss function!

It is also noted that Eq. (2.35) equals to 0 if and only if (IFF)

$$\frac{\partial L}{\partial \theta_i} = -\dot{\theta}_i = 0 \; \forall i \qquad (2.36)$$

That is, $\theta_i(t) = $ constant $\forall i$; which is the definition or condition for an equilibrium!

The above derivation is a formal proof that (continuous-time) gradient descent dynamics converge to local minima (or they remain at equilibria) only. Of course if the loss function is unbounded below, trajectories may go to $-\infty$ as stated above. That means, there are no other types of limit sets such as oscillations, limit cycles, or other exotic behaviors as chaos!

Now, to determine if an equilibrium is a min, max, or degenerate point, one computes the (symmetric square) *Hessian* matrix of the gradient system. The Hessian is the second partial derivative of the loss function evaluated at a point, say θ^*, i.e.,

$$H_{\theta*} := \left[\frac{\partial^2 L}{\partial \theta_i \partial \theta_j} \right]_{\theta*} \qquad (2.37)$$

where θ^* is an equilibrium point of interest.

Then from Calculus and Linear Algebra (Boyd & Vandenberghe, 2018; Riley et al., 2006), one infers the (local) property of an equilibrium point as summarized in Table 2.1.

For continuous-time gradient systems, the picture is simple. One can metaphorically envision a "landscape" of the loss as a function of all the parameters. A point or a ball may traverse this landscape starting from any point and move along paths defined by the gradient descent vector toward a local minimum point. It may be made to move at any speed on a continuous trajectory, and it would slow down as it approaches the "nearest" local minimum. End of story.

From the point of view of supervised learning in neural networks and deep learning (NNDL), however, the local minimum attained by the (continuous-time) gradient descent may still have a relatively large loss value, and thus not satisfactory for the final choice of parameters for a successful supervised learning (i.e., training and evaluation) phase.

Table 2.1 Equilibrium types

$H_{\theta*}$	Equilibrium	"Optimal" point
Positive definite	Asymptotically stable (sink)	Local minimum
Negative definite	Asymptotically unstable (source)	Local maximum
Nonsingular but non-definite	Isolated equilibrium (saddle)	Saddle
Singular	Degenerate point	Degenerate point

Again, in a crude way, one may view the local minima into two groups: (i) "good" minima, meaning the corresponding loss value is within the permissible performance level, and (ii) "bad" minima, meaning that the loss value is above the acceptable permissible performance level. The **NNDL** supervised learning will be successful, if the learning process phase ends with a "good" minimum; otherwise, it is not successful. Thus, the goal of the process in the learning phase is to attain a "good" parameter-space minimum that satisfies the permissible performance criterion (either satisfactorily low loss or satisfactorily high *evaluation* performance accuracy).

To that end, it appears that discretized dynamics of the update of the parameter could use large enough discretization step-sizes to allow the learning process to initially bypass (bad) local minima, and when appropriate to decrease the step-sizes sufficiently to follow the true (continuous-time) gradient descent to hone in on a (good) local minima. It is realized that this description is intuitive and learning techniques introduce numerous innovations toward realizing this intuition in practice, from infusing averaging of the gradients over mini-batches, to choosing variable learning rates that scale depending on the running root-mean-square (RMS) of the gradients or (the exponential of) the loss.

The computationally efficient discretized dynamics are a crude (i.e., first-order Euler) approximation to the continuous-time dynamics. To see that, we now approximate the time derivative of the parameter vector θ using first-order Euler approximation, i.e.,

$$\dot{\theta} \approx \frac{(\theta(t + \delta t) - \theta(t))}{(\delta t)} := \frac{\Delta \theta(t)}{(\delta t)} \tag{2.38}$$

Therefore, Eq. (2.29) becomes

$$\Delta \theta = - \delta t \, L_\theta := - \delta t \, \frac{\partial L}{\partial \theta} \tag{2.39}$$

The step-size δt plays a crucial control role. If the step-size is not "sufficiently small," then the discrete dynamics are no longer guaranteed to follow the (continuous-time) gradient systems dynamic properties. The dynamics may deviate markedly from the continuous-time dynamics and could exhibit more complex dynamics than just equilibria. As pointed out earlier, examples of these complex dynamics may range from limit cycles, oscillations, or even chaos!

Remark A2.1 If the gradient equations have a scalar factor γ, then the discretization would produce the term $\delta t \gamma$ as the crucial control factor often labeled in **NNDL** as the "learning-rate" η.

2.5   Appendix 2.2: The LMS Algorithm

We begin with the simple recurrent neural network (sRNN) to revisit the notation adopted in this book.

$$x_t = U h_{t-1} + W s_t + b \tag{2.40}$$

$$h_t = g(x_t) \tag{2.41}$$

$$z_t = V h_t + c \tag{2.42}$$

$$p_t = \psi(z_t) \tag{2.43}$$

Note that $\psi(\cdot)$ is typically the softmax function. At time step $t = 0$, this network requires the initial value h_{-1} and the input sequence $s_t = s_0, \cdots, s_n$ in order to generate the internal state sequence x_t (and the activation state sequence) and finally the output sequence z_t or p_t.

The LMS algorithm was initially developed for a single linear "neuron," which can easily be expanded to a multi-neuron (a vector), and one may add nonlinearity as well. It became popular in the engineering fields of signal processing and (linear) filters in the 1960s.

We shall describe it initially in the simplest form as a linear mapping of a (random) vector input to a (random) scalar output as follows:

$$x = w^T s = \sum_{k=1}^{k=m} w_k s_k \tag{2.44}$$

where the random variable vector s and the parameter vector w are m-dimensional vectors, and x is the random scalar output corresponding to the input s. One may view w^T as the first row of the matrix W in Eq. (2.40). We also use the notation for elements of the associated two vectors using the subscript index k.

Now, for each input vector s, the desired label is the scalar y. For the learning (i.e., training and evaluation) phase, one uses the loss function to be the L_2 norm squared of the error difference $y - x$. Then one applies the (stochastic) gradient descent to the loss function. The loss function is taken to be

$$J := \frac{1}{2} \mathbb{E}\{(y - x)^2\} \tag{2.45}$$

$$= \frac{1}{2} \mathbb{E}\{y^2 - 2xy + x^2\} \tag{2.46}$$

$$= \frac{1}{2} \mathbb{E}\{y^2\} - \mathbb{E}\left\{\sum_{k=1}^{k=m} w_k s_k y\right\} + \frac{1}{2} \mathbb{E}\left\{\sum_{j=1}^{j=m}\sum_{k=1}^{k=m} w_j w_k s_j s_k\right\} \qquad (2.47)$$

$$= \frac{1}{2} \mathbb{E}\{y^2\} - \sum_{k=1}^{k=m} w_k \, \mathbb{E}\{s_k y\} + \frac{1}{2} \sum_{j=1}^{j=m}\sum_{k=1}^{k=m} w_j w_k \, \mathbb{E}\{s_j s_k\} \qquad (2.48)$$

where $\mathbb{E}$ is the expectation operator. We used the following steps above: (i) we expanded the top expression, (ii) used the definition of the linear network, (iii) used the fact that the expectation operator $\mathbb{E}$ is linear and thus passes through summations, and (iv) exploited the assumption that the parameters (weights) are deterministic variables.

Thus, the loss function is now a function of:

(i) The square-mean-value of the labels, namely, $\mathbb{E}\{y^2\}$
(ii) The cross-correlation functions, $\mathbb{E}\{s_k y\}$ between the input elements and the label
(iii) The auto-correlation functions $\mathbb{E}\{s_j s_k\}$ among the elements of the input signal

Next, one obtains the optimal parameter w. We calculate the gradient of J with respect to each element of w.

$$\nabla_{w_k} J := \frac{\partial J}{\partial w_k}$$

$$= -\mathbb{E}\{s_k y\} + \sum_{j=1}^{j=m} w_j \, \mathbb{E}\{s_j s_k\}, \quad \forall k \qquad (2.49)$$

We also note that the second derivative of J is the Hessian and can be obtained as follows:

$$\nabla_{w_k w_j} J := \frac{\partial^2 J}{\partial w_k, \partial w_j}$$

$$= \mathbb{E}\{s_j s_k\}, \quad \forall k, j \qquad (2.50)$$

Now, there are two approaches in finding the optimal parameters of this one neuron example (or linear filter):

Approach 1 Set the gradient to zero, and then solve for the optimal parameters. The second derivative of J provides the Hessian matrix that would determine if the optimal point is stable (sink), unstable (source), saddle, or degenerate point.

In this context, this linear filter gradient system is in fact a **convex optimization problem** (Boyd & Vandenberghe, 2004, 2018). Thus, at optimal weights (min, max, or saddle), one has

$$0 := -\mathbb{E}\{s_k y\} + \sum_{j=1}^{j=m} w_j\, \mathbb{E}\{s_j s_k\}\quad \forall k \tag{2.51}$$

Collecting these sets of m-equations in a vector-matrix form, one obtains the equality:

$$[0] := -\,\mathbb{E}\{sy\} + \mathbb{E}\{ss^T\}w \tag{2.52}$$

where we use [0] to denote the m-d vector of zeros. Or in this form as a linear vector-matrix equation in the unknown parameter vector w.

$$\mathbb{E}\{sy\} = \mathbb{E}\{ss^T\}w \tag{2.53}$$

where $\mathbb{E}\{ss^T\}$ is the *symmetric* covariance matrix (with elements $\mathbb{E}\{s_j s_k\}$), and $\mathbb{E}\{sy\}$ is the cross-variance of the input m-d vector s with the desired (scalar) label y (with elements $\mathbb{E}\{s_k y\}$). One observes that Eq. (2.53) is the equation for the Wiener optimal filter, which, if solved, would produce the optimal (Wiener) filter parameter vector w.

Solution of the Wiener Filter
The (global) unique solution to the Wiener filter requires that the (symmetric) covariance matrix be nonsingular (and thus invertible). Then a unique optimal filter parameter vector w^o will be obtained as

$$w^o = \mathbb{E}\{ss^T\}^{-1}\,\mathbb{E}\{sy\} \tag{2.54}$$

This solution provides then the global minimum if the (nonsingular) covariance is also positive definite. Symmetric matrices have only real eigenvalues, and thus if all the eigenvalues of the covariance are positive, then the covariance is positive definite.

There are two challenges to this approach:

(1) Estimating the true cross- and auto-correlation without knowledge of the probability densities of the random variables (here the input vector s and the label scalar y).
(2) Computing an inverse of a matrix. This may be an easy computational task for low-dimensional cases, and it becomes challenging as the dimensions increase.

Approach 2: The Method of Steepest Descent (Stochastic Gradient Descent)
One defines the gradient dynamics for the parameters incrementally as

$$\Delta w_t := -\eta \frac{\partial J}{\partial w} = \mathbb{E}\{sy\} - \mathbb{E}\{ss^T\}w_t \tag{2.55}$$

where we inserted the iteration (time) index t. Iterating this gradient system should lead to the global optimal solution of this convex optimization linear system with a quadratic loss function.

This approach eliminated the need for computing the inverse. However, the challenge of estimating the expected value of the cross-correlation and auto-correlation still remains.

The Adaptive Least-Mean-Square (LMS) Algorithm
(1) An assumption: Let the input and the corresponding label come from an *ergodic* process. Thus the statistical ensembles can be estimated by the corresponding time averaging.
(2) This allows using samples over time. Thus one may remove $\mathbb{E}(\cdot)$ from the dynamic process of the (stochastic) gradient descent and settle for time iterations of the dynamics of the parameter vector. This sometimes also referred to as *on-line learning*.

Thus, we relax the equation of (2.55) by removing the expectation operator $\mathbb{E}$ to obtain the estimated change in the update weight. Thus, the LMS update equations are as follows:

$$\Delta \hat{w}_t := -\eta \frac{\partial J}{\partial \hat{w}_t} \tag{2.56}$$

$$\approx \eta sy - ss^T \hat{w}_t \quad t \geq 0 \tag{2.57}$$

$$= \eta s(y - x) \tag{2.58}$$

which gives the parameter-update equation as

$$\hat{w}_{t+1} = \hat{w}_t + \Delta \hat{w}_t \quad t \geq 0 \tag{2.59}$$

$$= \hat{w}_t + \eta s(y - x), \quad t \geq 0 \tag{2.60}$$

Observe that the weight equation for a row (instead of a column) parameter vector becomes

$$\hat{w}_{t+1}^T = \hat{w}_t^T + \eta(y - x)s^T, \quad t \geq 0 \tag{2.61}$$

Thus, the equation for the parameter matrix $\hat{W}$, when one has multiple outputs forming the n-d vector x, can be easily obtained (or similarly derived) to be

$$\hat{W}_{t+1} = \hat{W}_t + \eta(y - x)s^T, \quad t \geq 0 \tag{2.62}$$

where now the difference $(y - x)$ is an n-d vector.

The LMS Algorithm Procedure

The procedure follows the list of *actions*:

(1) Initialization: Choose the hyper-parameter (learning rate) η. Initialize the weight matrix $\hat{W}_0$ with small random element values. (Choose the learning rate initially relatively large, and then decay its value (e.g., exponentially) over time.)

(2) Forward pass: Pick a sample input s, and run the network (forward) to obtain the corresponding output signal vector x.

(3) Backward pass: Compute the error vector $y - x$. (In this single layer example, there is no network to pass through backward. This will happen in the multi-layer or recurrent neural network cases as we will see in the next chapter—Chap. 3.)

(4) Update the weight matrix one step as in Eq. (2.62). (Or repeat actions (2) and (3) and update over several steps or a **mini-batch**).

(5) Iterate. Go to action (2) for either a pre-specified number of **epochs** or until the (L_2 norm of the) error vector $y - x$ becomes (and stays) below an acceptable level.

Exercise A1

Show that for the one-layer feedforward neural network given as:

$$x_t = W s_t + b \tag{2.63}$$

$$h_t = g(x_t) \tag{2.64}$$

The weight W update law using the *on-line* version of the LMS is

$$W_{t+1} = W_t + \eta \{g'\}(y - x)s^T, \quad t \geq 0 \tag{2.65}$$

$$b_{t+1} = b_t + \eta \{g'\}(y - x), \quad t \geq 0 \tag{2.66}$$

where $\{g'\}$ is the diagonal matrix formed of the derivative of the vector of nonlinearities $g(\cdot)$. (Or $\{g'\}$ is simply the derivative vector if one uses the point-wise (Hadamard) multiplication.) More generally, the updates can be expressed as

$$W_{t+1} = W_t + \eta \{g'\} \mathbb{E}\{(y - x)s^T\}, \quad t \geq 0 \tag{2.67}$$

$$b_{t+1} = b_t + \eta \mathbb{E}\{(y - x)\}, \quad t \geq 0 \tag{2.68}$$

where $\mathbb{E}$ is the expectation operator that can be approximated by a single or multiple (i.e., mini-batch) measurements or samples.

Part II
Recurrent Neural Networks (RNN)

Summary Part II is composed of only one chapter: Chap. 3. It includes the fundamental technical material on recurrent neural networks (RNN). It casts the supervised learning task in the principled form of Adaptive Non-convex Optimization under dynamic network constraints and leads to the *generalized* form of the backpropagation through time (BPTT). This holistic treatment brings systemic depth as well as ease to the process of adaptive learning for general (i.e., gated and ungated) recurrent neural networks.

Chapter 3
Recurrent Neural Networks (RNN)

3.1 Simple Recurrent Neural Networks (sRNN)

The **simple recurrent neural network** (sRNN) can be viewed as a single layer recurrent neural network where the activation is delayed and fed back synchronously with the external input (or a previous layer's output). Mathematically, the simple recurrent neural network (sRNN) is expressed as

$$h_t = \sigma_t(Uh_{t-1} + Ws_t + b), \qquad t = 0, \cdots, N \qquad (3.1)$$

where t is the discrete, real or fictitious, time index, N is the **finite-horizon** final time, s_t is the m-d external input vector, and h_t is the n-d output **activation** via the nonlinear function σ_t. Here, σ_t is a general, possibly time-varying, nonlinear function. However, it is typically specified to be the logistic function or the hyperbolic tangent *tanh* as is common in the literature (see Bengio et al., 2015; Jianxin et al., 2016, and the references therein), or even as the **rectified linear unit** reLU as in Jaitly et al. (2015). The non-indexed parameters, to be determined by training, are the $n \times n$ matrix U, the $n \times m$ matrix W, and the $n \times 1$ vector bias b.

This model is a discrete nonlinear dynamic system, with finite-horizon time steps, see, e.g., Graves (2012) and Chung et al. (2014a). Recurrency (or feedback) is expressed by delaying the output activation function h_t, transforming it by the matrix U, and feeding it through the nonlinearity $\sigma_t(\cdot)$ in Eq. (3.1).

Remark 3.1 From a **system and signal processing** viewpoints, the model in Eq. (3.1), without the nonlinearity (σ_t), represents a linear filtering that can be expressed in the (frequency) Z-transform as follows:

$$H(z) = \left[I - Uz^{-1}\right]^{-1}(WS(z) + b)$$

$$= z\,[zI - U]^{-1}(WS(z) + b)$$

© The Author(s), under exclusive license to Springer Nature Switzerland AG 2022
F. M. Salem, *Recurrent Neural Networks*,
https://doi.org/10.1007/978-3-030-89929-5_3

In general, this represents a high-pass filtering, and thus all high frequency input components, including noise, will pass through to the output.

We can now re-write this sRNN architecture as

$$x_t = U h_{t-1} + W s_t + b, \qquad t = 0, \cdots, N \qquad (3.2)$$

$$h_t = \sigma_t(x_t), \qquad t = 0, \cdots, N \qquad (3.3)$$

where we explicitly identify the internal state vector x_t. Equation (3.2) captures the dynamic and stability behavior; and Eq. (3.3) is a static (non-dynamic) nonlinear mapping of the state to the activation (state) h_t. The (local) stability properties of the network are determined by the matrix $U h'_{t-1}$, where h'_{t-1} represents the diagonal matrix of the derivative of the vector h_{t-1} at the state x_{t-1} value. However, the matrix U is typically initialized to be randomly small and is adaptively changing over the training iterations. That is, $U h'_{t-1}$ may exhibit or lead to "unstable" behavior of the dynamic network during the training phase! If $U h'_{t-1}$ is an unstable matrix (for a simple special case, this may mean that at least one of its eigenvalues has a modulus greater than unity) at the end of training, the nonlinear dynamic system *may* become unstable, which may lead to x_t growing without bound. Even though the activation h_t is bounded when using a compressive nonlinearity for σ_t, the state x_t can become unbounded and thus the dynamic system may in fact become unstable in the sense that x_t grows unbounded. Of course, Liapunov theory and Liapunov functions would constitute a definitive approach to confirming the stability of the overall nonlinear network during or after training.

We further add subsequent static linear layer (but without external input) as

$$z_t = V h_t + c, \qquad t = 0, \cdots, N \qquad (3.4)$$

$$p_t = g_t(z_t), \qquad t = 0, \cdots, N \qquad (3.5)$$

where, as before, $t = 0, \cdots, N$ is the discrete (time) index, h_t is the activation n-d vector, sometimes called the hidden unit, resulting from applying the nonlinear function $\sigma_t(.)$ to the state vector x_t. The r-d vector z_t is the consequent output. Equation (3.5) applies the nonlinearity $g_t(\cdot)$—typically the softmax function is used—to convert the linear layer output signal vector z_t into an estimated output probability vector p_t.

The sRNN architecture, expressed by Eqs. (3.2–3.5), can be viewed as standard. It is observed that (i) state x_t represents the information (or information memory) of the input sequence up to, and including the instant t, and (ii) there is a single pathway from the input s_t to the output z_t. These two observations will be referred to in Sect. 3.2.

For training this architecture, one has to choose a method to determine the parameters for an application of interest. The available methods today use algorithms to adaptively (and incrementally) change the parameters starting from initial (typically, random) guesses. The adopted dominate methods for training have been families, or versions, of the stochastic gradient descent (SGD). Why SGD? First, it has been found to do a good job in practice in training neural networks in general. Second, the ingenuity efforts have been focusing on innovative designs of network architectures while keeping the training methods within the families of SGD. Third, while SGD requires sufficiently large amounts of training data for high accuracy performance, this is no longer a problem due to the abundance of data for many applications. Fourth, the advent of high-speed and parallel computing, think of GPUs and cloud computing, algorithmic executions can be achieved in reasonable times. While many deep learning frameworks, e.g., TensorFlow and Keras, PyTorch, etc., hide the training methods "under_the_hood," it is important to understand the details for technical understanding of the methods as, more importantly, to perhaps have to extend or innovate such methods in the future.

3.1.1 The (Stochastic) Gradient Descent for sRNN: The Backpropagation Through Time (BPTT))

We will pursue a walk-through of how to obtain the parameter-update equations using **backpropagation through time** (BPTT). The BPTT is simply a recursive computation of the stochastic gradient descent using the derivative chain rule in basic calculus, see, e.g., Graves (2012).

First, one chooses a loss function $L(\cdot)$ that may be a metric (or a measure) defined over the finite-horizon time sequence $t = 0, \cdots, N$ as

$$L := \mathbb{E}\{L^t\} \approx \frac{1}{N'} \sum_{t=0}^{N} L^t \tag{3.6}$$

where $\mathbb{E}$ is the expectation operator of the finite-horizon sequence. As the probabilities are unknowns, one typically assumes a uniform probability of $\frac{1}{N'}$, where N' equals $N + 1$. In optimizing the loss function, however, N''s value is not important as it would be absorbed into the parameter adaptive update *learning rate*.

We seek to compute the gradient expressions of this loss function with respect to each of the parameters in preparation for estimating the (stochastic) gradient descent for updating such parameters. To that end, we express the standard gradient descent for a generic parameter, say θ, as

$$\Delta\theta := -\eta \frac{\partial L}{\partial \theta} \tag{3.7}$$

where $\Delta\theta$ expresses the simple (Euler) approximation of the time derivative, η is the effective learning rate, often called a **hyper-parameter**, to signify that such (hyper-)parameters are to be set by the user or determined by alternate methods (including slower time-scale SGDs).

We now detail the expressions for the gradients for each of the specific realization of the parameter θ in the sRNN. Note that the factor $\frac{1}{N'}$ is absorbed into the *learning rate* η:

(i) In the output equation, first, we define the column vector equality:

$$e_t := \left(\frac{\partial L^t}{\partial z_t}\right) \tag{3.8}$$

Then, we get

$$\frac{\partial L}{\partial V} = \sum_{t=o}^{N} \left(\frac{\partial L^t}{\partial z_t}\right) \frac{\partial z_t}{\partial V}$$

$$= \sum_{t=o}^{N} e_t h_t^T \tag{3.9}$$

It is noted that the size of this partial derivative is the same as the size of the matrix V, or specifically the change matrix ΔV.

Analogously, for the change Δc we have

$$\left(\frac{\partial L}{\partial c}\right)^T = \sum_{t=o}^{N} \left(\frac{\partial L^t}{\partial z_t}\right)^T \frac{\partial z_t}{\partial c}$$

$$= \sum_{t=o}^{N} (e_t)^T \tag{3.10}$$

where it is noted that the last partial derivative with respect to c is equal to the identity. Moreover, we used the transpose in Eq. (3.10) to retain chain-rule derivative flow from the scaler case in calculus.

(ii) For the parameters in the dynamic equation, first we define the intermediate **sensitivity** variable λ by the column vector equality

$$\lambda_t = \left(\frac{\partial L^t}{\partial x_t}\right) \tag{3.11}$$

which can be expressed based on variation principles as a backward-in-time dynamic equation in the following:

$$\lambda_t = (U_k \sigma_t')^T \lambda_{t+1} + (V_k \sigma_t')^T e_t$$

$$0 < t \leq N - 1 \tag{3.12}$$

Remark 3.2 Equation (3.12) can be derived from the fact that the variable x_t forwardly connects to the output at time t and the next time $t+1$. More precisely, the derivation is obtained from the variation principles as presented in the next section, see Eq. (3.44). Note also that the parameters (i.e., the matrices) are indexed with their own iteration index k over their mini-batch updates.

Then, we get the following expression using the chain rule:

$$\frac{\partial L}{\partial U} = \sum_{t=o}^{N} \left(\frac{\partial L^t}{\partial x_t} \right) \frac{\partial x_t}{\partial U}$$

$$= \sum_{t=o}^{N} \lambda_t \, h_{t-1}^T \tag{3.13}$$

The (sensitivity) variable λ_t in some references is replaced by δ_t and is referred to as the delta variable in the backpropagation through time (BPTT). We choose to express it as λ_t to be consistent with the labeling in the **Lagrange multiplier** and the **Calculus of Variations** in Optimization Theory from which these computations originate. This choice is consistent with the subsequent developments.

Similarly, we calculate the partial derivatives for the remaining parameters.

$$\frac{\partial L}{\partial W} = \sum_{t=0}^{N} \left(\frac{\partial L^t}{\partial x_t} \right) \frac{\partial x_t}{\partial W}$$

$$= \sum_{t=0}^{N} \lambda_t s_t^T \tag{3.14}$$

$$\left(\frac{\partial L}{\partial b} \right)^T = \sum_{t=o}^{N} \left(\frac{\partial L^t}{\partial x_t} \right)^T \frac{\partial x_t}{\partial b}$$

$$= \sum_{t=o}^{N} (\lambda_t)^T \tag{3.15}$$

The sRNN dynamic equations and the associated BPTT algorithm can be depicted and visualized in a diagram sketch as in Fig. 3.1. (To simplify the busy diagram, we assume V is the identity matrix and $c = 0$), and then e_t is the instantaneous error at each time instant over the finite horizon.

Now, using variants of the stochastic gradient descent to train sRNN for long-term dependencies or sequences, it has been widely reported that the sRNN— in computational practice—may exhibit the so-called vanishing gradient or the exploding gradient, Graves (2012), Chung et al. (2014a) and Jaitly et al. (2015), and the references therein. Thus, it has been largely accepted that sRNNs are difficult to train by the stochastic gradient descent, see Graves (2012), Bengio et al. (2015), Chung et al. (2014a) and the references therein.

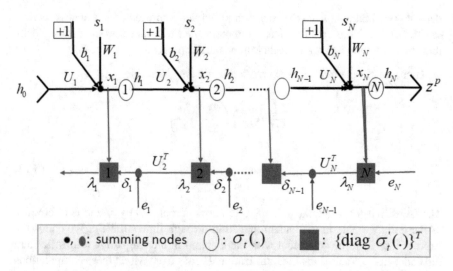

Fig. 3.1 Visualization diagram of the BPTT for sRNN

This has spurred developments of new training approaches, modifications, and more complex architectures to enable convergent training and learning in state-of-the-art applications (Chung et al., 2014a; Graves, 2012; Jianxin et al., 2016; Zaremba et al., 2014). Recent trends have focused on reducing the computational load and complexity while preserving their demonstrated learning capabilities of sequences and time-series (Graves, 2012; Jaitly et al., 2015; Zaremba et al., 2014).

The so-called exploding gradient phenomena—whereby the calculated gradient grows unbounded over the "time" index t—has been addressed in simulation practice by **clipping** the computed gradients when their amplitudes exceed a pre-specified upper bound. While this would permit the gradient descent procedure to continue on, it does not necessarily lead to convergent behavior during training.

There has been interest in reviving simple RNNs with modifications in the type of nonlinearity and an approach toward careful training initialization. For example, the introduction of the so-called **IRRN design** that uses the sRNN with rectified Linear Units (ReLUs) for the nonlinearity, and an initialization of the U matrix (in Eq. (3.2) to be the identity matrix, and setting a cap limit on the resulting gradient computation. Such approach is introduced in Jaitly et al. (2015).

3.2 Basic Recurrent Neural Networks (bRNN)—the General Case

We describe here simple modifications that would enable the sRNN to overcome its limitations to successful training. To begin with, it would be beneficial to add a (stable) linear (affine) term in the state equation (3.2) that bounds the growth of the state x_t to within a bounded region. In Appendix 3.1, we include the analysis based on the Liapunov theory to show that such dynamic system would indeed be bounded to a region in the (x)-state space.

In addition, one should include a subsequent linear output layer in the modeling—with direct path connection from the input signal—to allow for flexible filtering that is intended to render the recurrent network with its output layer a general *bandpass* filter. Such a filter would have the equivalent of (nonlinear) poles and zeros that, after training, could produce a linearizable system with an equivalent *proper* stable transfer function. This motivates one to also revise the indexing at the input. This description serves as a motivation of the simplest redesign of the nonlinear architecture.

Let us define a **basic recurrent neural network** (bRNN) as an architecture that can achieve **regression** (e.g., output tracking) of real (i.e., analog) output values. We shall use the classical approach for **non-convex constrained optimization**. This is the tool that explains the gradient descent approach and its backpropagation through time manifestation. We seek to obtain a **bounded input-bounded-state** (BIBS) stable flexible recurrent neural network that can be trained to learn (discrete) categories or (continuous) regression profiles. For ease of presentation, we present the overall architecture (or model) and relegate detailed derivation to Appendix 3.2 at the end of this chapter.

This bRNN has the following unique attributes:

- Stable behavior without a need of additional "gating" networks. Gating networks would at least double the number of the overall adaptive parameters in the overall RNN. In addition, no need for special caps on the growth of the gradient signals.
- Specify a predictive state space formulation. Specifically, the network takes input data and state values at one index (or time) value, say t, and produces a state value and/or output at the next index, say $t + 1$. In this formulation, the state x_t serves as an information memory of all prior values of the input sequence, i.e., s_i, $i = 0, \ldots, t - 1$, while the present input s_t brings the new input instance not contained in the memory state x_t. This is in contrast to the classical sRNN where the state contains the present input s_t.
- Update law derivations, including the backpropagation network, are shown to follow easily using the Lagrange multiplier constrained optimization method. One identifies the forward and backward propagation networks and the split-boundary conditions in a principled way. The split-boundary conditions are the initial state and the final co-state, and they are the fundamental reason behind the state-forward and co-state-backward propagating dynamic processing for the stochastic gradient descent.
- In this framework, extensions and definitions of the output or state loss functions can easily be incorporated without complexity. For example, each loss function can be independently specified to be a supervised loss (with a given label/target) or an unsupervised loss function, e.g., minimizing the **entropy** or maximizing **sparsity** in an internal representation (state) or hidden activation vector (Waheed & Salem, 2003).

We describe the basic recurrent neural network (or bRNN) that includes a separate linear (i.e., affine) term with a slightly "stable" fixed matrix to guarantee bounded solutions and fast dynamic response. We formulate a state space viewpoint and adapt the constrained optimization Lagrange multiplier technique and the vector **Calculus of Variations** (CoV) to derive the (stochastic) gradient descent. In this process, one avoids the commonly used re-application of the circular chain rule and identifies the error backpropagation with the **co-state**-backward dynamic equations. The bRNN can successfully perform regression, tracking, or prediction of time-series. Moreover, the "vanishing and exploding" gradients are explicitly quantified and can be explained through the co-state dynamics and the update laws. The adapted CoV framework, in addition, can correctly and principally integrate new loss functions in the network on any variable and for varied goals, e.g., for supervised learning on the outputs and unsupervised learning on the internal (hidden) states.

In contrast to the sRNN, see Eqs. (3.2)–(3.3), now consider this extended recurrent neural network model:

$$x_{t+1} = Ax_t + Uh_t + Ws_t + b, \qquad t = 0, \cdots, N-1 \qquad (3.16)$$

$$h_t = \sigma_t(x_t), \qquad t = 0, \cdots, N \qquad (3.17)$$

$$z_t = Vh_t + Ds_t + c, \qquad t = 0, \cdots, N \qquad (3.18)$$

$$p_t = g_t(z_t), \qquad t = 0, \cdots, N \qquad (3.19)$$

where, as before, $t = 0, \cdots, N$ is the discrete (time) index, s_t is the m-d input vector sequence, x_t is the n-d state vector, and h_t is the activation n-d vector, sometimes labeled as the hidden unit, resulting from applying the nonlinear function $\sigma_t(.)$ to the state vector x_t. The r-d vector z_t is the subsequent output layer sequence. Equation (3.19) applies the nonlinearity $g_t(.)$ to convert the linear output signal z_t into an estimated output probability vector p_t.

For this neural system, Eq. (3.16) includes the dynamic transition from the index step t (all variables on the right-hand side have the same index t) to the state at the next step $t + 1$. The state vector x_t depends on the prior elements of the input sequence up to $t - 1$. This is distinctly different from the definition of the state in sRNN which depends on the input sequences up to the present time t! Eq. (3.17) represents the static nonlinear state transformation that may include any of the common nonlinearities, e.g., the logistic function, hyperbolic tangent, reLU, etc. Equation (3.18) is the output (linear) layer that now includes a direct input path from the instantaneous input $s(t)$. Equation (3.19) applies the nonlinearity $g_t(.)$, which is often a *logistic* or *softmax* function.

Remark 3.3 To keep with practice, specially in line with the computational platforms and codes, Eq. (3.16–3.17) may be re-written as

$$h_o = \sigma_o(x_o),$$

$$x_t = Ax_{t-1} + Uh_{t-1} + Ws_{t-1} + b, \qquad t = 1, \cdots, N \qquad (3.20)$$

$$h_t = \sigma_t(x_t), \qquad\qquad\qquad\qquad t = 1, \cdots, N \qquad (3.21)$$

where the initial state x_o and the input sequence s_t, $t = 0, \cdots, N - 1$ need to
be available to recursively generate the state sequence x_t, $t = 1, \cdots, N$, and the
"activation" sequence h_t, $t = 0, \cdots, N$. However, without loss of generality, we
will adopt the form of Eq. (3.16) for the derivation development of the gradient
descent for convenience as one can easily "shift" the sequence indexing as is done
here.

This neural model extends the simple recurrent neural networks (sRNN) by
adding *a linear or affine* state term in the dynamic equation, Eq. (3.16), via the
matrix A, and also by adding *the direct input* term in the output equation, Eq. (3.18),
via the matrix D. All terms on the left side of the equations exhibit the same time
index t to predict the state at the time index $t + 1$. The dimensions of the parameters
should be obvious to achieve compatibility of the equations. Specifically, A and U
are $n \times n$, W is $n \times m$, b is $n \times 1$, V is $r \times n$, D is $r \times m$, and c is $r \times 1$.

In Eq. (3.16), the matrix A is set to be constant with eigenvalues having moduli
within (but near) the unit circle. As a special case, one may choose the eigenvalues
to be distinct, real, or random, with amplitude ≤ 1. For large scale models, it may
be computationally easier to choose $A = \alpha I$, for $0 < \alpha \leq 1$. The matrix A must be
a *stable* matrix. In Eq. (3.18), the direct input term enriches the model since the state
equation can only produce transformations of delayed versions of the input signal
sequence but not an instantaneous input.

The parameters in the dynamic equation, Eq. (3.16), enumerated in the matrices
U, W, and the bias vector b, can be represented by a single vector parameter θ,
whereas the parameters of the output equation (3.18), enumerated in the matrices
V, D, and the bias vector c, can be represented by the single vector parameter v. We
shall also index these parameters by the index t in order to facilitate the upcoming
discussion. Thus we shall represent the dynamic equations as follows:

$$x_{t+1} = Ax_t + U_t h_t + W_t s_t + b_t, \qquad t = 0, \cdots, N - 1$$

$$=: f^t(x_t, h_t, s_t, \theta_t)$$

$$h_t = \sigma_t(x_t), \qquad\qquad\qquad\qquad t = 0, \cdots, N$$

$$z_t = V_t h_t + D_t s_t + c_t, \qquad\qquad t = 0, \cdots, N \qquad (3.22)$$

$$p_t = g_t(z_t), \qquad\qquad\qquad\qquad t = 0, \cdots, N$$

$$=: g^t(h_t, s_t, v_t)$$

where s_t, $t = 0, \cdots, N$ is the sequence of input signal, and t is the time index of the dynamic network. In the case of supervised adaptive learning, we associate with each sample sequence s_t the desired label sequence y_t.

The cost or *loss* function is thus given in general as a function of the final-time output as well as the variables at the intermediate times within $o \leq t < N$. Here, for less cumbersome calculations, we choose the final-time output to be z_N and define the loss function over the finite-time horizon to be

$$J_o(\theta_o, \ldots, \theta_{N-1}, v_o, \cdots, v_N) = \phi(z_N) + L^N(v_N)$$

$$+ \sum_{t=o}^{N-1} L^t(z_t, x_t, h_t, \theta_t, v_t)$$

where this general loss function includes all network variables and parameters.

Remark 3.4 In this development, we shall compare the desired *labels* with the output sequence z_t, not p_t, in order to make the description and the discussion easier. It would involve simple insertions in the calculations if the labels were used for the sequence p_t instead. We leave that process as an exercise for the attentive reader.

Following the **Calculus of Variations** (CoV) in constrained optimization and the **Lagrange multiplier** technique, see e.g., Bryson and Ho (1975) and Vrabie et al. (2012), one would define the Hamiltonian at each time step t as

$$H^t = L^t(z_t, x_t, h_t, \theta_t, v_t) + (\lambda_{t+1})^T f^t(x_t, h_t, s_t, \theta_t),$$

$$t = 1, \cdots, N - 1$$

where the sequence λ_t, $0 \leq t \leq N$, of the Lagrange multipliers would become the co-state with dynamics generated as in constrained optimization and optimal control, see Bryson and Ho (1975) and Vrabie et al. (2012). The state equations and the **co-state** equations are generated from optimization, similar to optimal estimation and control procedures. We then apply the (stochastic) gradient descent to all the parameters. Recall that the parameter vector θ_t represents the elements of the matrices U_t, W_t and bias vector b_t. Similarly, the parameter vector v_t represents the elements of the matrices V_t, D_t and the bias vector c_t.

Thus, the (vector) state equation representing the network is reproduced as

$$x_{t+1} = \frac{\partial H^t}{\partial \lambda_{t+1}} = f^t$$

$$= Ax_t + U_t h_t + W_t s_t + b_t, \qquad t = 0, \cdots, N - 1 \qquad (3.23)$$

where the output layer is

$$z_t = V_t h_t + D_t s_t + c_t, \qquad t = 0, \cdots, N \qquad (3.24)$$

The **co-state** dynamics are generated as

$$\lambda_t = \frac{\partial H^t}{\partial x_t}$$

$$= \frac{\partial L^t}{\partial x_t} + \left(\frac{\partial f^t}{\partial x_t}\right)^T \lambda_{t+1}, \qquad t = 0, \cdots, N \qquad (3.25)$$

The co-state dynamics are basically linearized (**sensitivity**) equations along the state trajectory x_t, $t = 0, \cdots, N$.

The gradient descent change in the parameters (within the state equation) at each time step t is

$$\Delta \theta_t = -\eta \frac{\partial H^t}{\partial \theta_t}$$

$$= -\eta \left(\frac{\partial L^t}{\partial \theta_t} + \left(\frac{\partial f^t}{\partial \theta_t}\right)^T \lambda_{t+1}\right) \qquad (3.26)$$

where η is a general (sufficiently small) learning rate. Similarly, the gradient change in the parameters (within the output equation) at the time step t is

$$\Delta \nu_t = -\eta \frac{\partial H^t}{\partial \nu_t} = -\eta \left(\frac{\partial L^t}{\partial \nu_t}\right) \qquad (3.27)$$

Note that Eqs. (3.26)–(3.27) are written as deterministic expressions for clarity; however, they should be viewed with an expectation operator applied to the right-hand side. These changes are expressed at each time step t; thus, there is a "sequence" of gradient changes over the whole (finite-)time horizon duration $[0, \cdots, N]$. The goal is to use these time-step parameter changes to calculate the changes to the parameters over the whole time horizon due to an example trajectory sequence. The sequence is considered a "*sample*" in stochastic training. We note that the parameters of the network are "*shared*" over the time horizon $[0, \cdots, N]$ and assumed to be constant during the sample trajectory. They can only be incremented (or updated) from a *mini-batch* to the next *mini-batch* during training–till they converge to (computationally near) constant values with small parameter variations, or better yet, a loss value within a small acceptable bound.

Finally, the **split boundary conditions** must always be satisfied and are crucial to the optimization procedure. The technically derived split boundary conditions are:

$$\left(\frac{\partial \phi}{\partial x_N} - \lambda_N\right)^T dx_N = 0 \qquad (3.28)$$

$$(\lambda_o)^T dx_o = 0 \qquad (3.29)$$

In recurrent neural networks, it is usually assumed that the initial state stage x_o is given or specified as constant, and thus $dx_o = 0$. Therefore, Eq. (3.29) is trivially satisfied. Also, for the boundary condition Eq. (3.28), we make the quantity in parenthesis to (always) equal zero. The (split) boundary conditions thus become:

(i) Initial boundary condition:

$$x_o = \text{constant} \tag{3.30}$$

(ii) Final boundary condition:

$$\lambda_N = \frac{\partial \phi}{\partial x_N} \tag{3.31}$$

Thus, given a set of parameters' values, the (training) iteration process, per *sample*, is to use the (same) initial state x_o and apply the sample's input signal sequence s_t and the corresponding desired sequence y_t to generate the state and output sequences (forward propagation in time) up to the final time N. Then, compute the final boundary of the **co-state** using Eq. (3.31) and subsequently generate the co-state sequence λ_t of Eq. (3.25) backward propagation in time. At the end of each (forward–backward) iteration, or a batch of iterations (namely, a **mini-batch**), all parameters would then be updated and continue doing so according to some **stopping criterion**. In principle, this is the core parameters' iteration procedure in training neural networks using SGD. After each iteration (or mini-batch) is completed, and/or after the stopping criterion is met, one may use the achieved fixed parameters in the recurrent network, Eqs. (3.23)–(3.24), for evaluation, testing, or processing—i.e., **inference**.

Remark 3.5 We want to take a special note here of the fact that the adaptive optimization under the network dynamic constraints is applicable to broad network architectures as expressed generally with a function $f^t(\cdot)$ in Eqs. (3.22) and the subsequent equations. In particular, these derivations are applicable to the **gated RNNs** that will be covered in the next Parts III–IV.

3.3   Basic Recurrent Neural Networks (bRNN)—a Special Case

We now provide specific details of the bRNN architecture and its parameter updates. We basically specify the components of the loss function and the boundary conditions, i.e., the initial condition for the state-forward dynamics and the final condition for the co-state-backward dynamics.

The final-time component of the loss function, namely $\phi(z_N)$, can be specified for supervised learning as, e.g., an L_2 norm (squared) based error quantity at the final-time step N. Specifically,

$$\phi(z_N) = \frac{1}{2}||z_N - y_N||_2^2$$

$$= \frac{1}{2}(z_N - y_N)^T (z_N - y_N) \qquad (3.32)$$

And thus its derivative is calculated from **vector calculus** to produce the final condition of the co-state as

$$\lambda_N = \frac{\partial \phi}{\partial x_N} = \left(\frac{\partial z_N}{\partial x_N}\right)^T \frac{\partial \phi}{\partial z_N} \qquad (3.33)$$

$$= (\sigma_N')^T V_N^T (z_N - y_N) \qquad (3.34)$$

where one uses the output equation (3.24) for the definition of z_N. The matrix σ_N' is the derivative of the nonlinearity vector σ_N expressed as a diagonal matrix of element-wise derivatives. This provides the final co-state at the final time N.

Remark 3.6 Note that for computational expediency, the diagonal matrix σ_N' can be expressed as a vector of derivatives multiplied point-wise (i.e., a **Hadamard multiplication**) to the vector $V_N^T (z_N - y_N)$ on the right or the left. This is more commonly adopted in computational/coding implementations.

A more popular loss function in current practice is the *cross-entropy* loss, i.e.,

$$\phi(z_N) = -y_N{}^T \log(z_N) = - \sum_{i=o}^{i=r} y_{Ni} \log(z_{Ni}) \qquad (3.35)$$

Its derivative is calculated from **vector calculus** to produce the final condition of the co-state as

$$\lambda_N = \frac{\partial \phi}{\partial x_N} = \left(\frac{\partial z_N}{\partial x_N}\right)^T \frac{\partial \phi}{\partial z_N} \qquad (3.36)$$

$$= - (\sigma_N')^T V_N^T [(y_{N1}/z_{N1}) \ldots (y_{Nn}/z_{Nn})]^T \qquad (3.37)$$

where we used Eq. (3.33) for the chain rule vector calculus.

To ease the following computations, the general loss function may be simplified to be a sum of separate loss functions in each vector variable, e.g., as

$$L^t(z_t, x_t, h_t, \theta_t, v_t) = L^t(z_t)$$

$$+ \beta_x L^t(x_t) + \beta_h L^t(h_t)$$

$$+ \gamma_1 L^t(\theta_t) + \gamma_2 L^t(v_t) \qquad (3.38)$$

where the first term is a component loss function on the output at index t and is chosen to be a supervised loss function with corresponding reference or target y_t. One may use the scaled L_2 norm squared of the error to be consistent with the final-time loss function in Eq. (3.32), specifically,

$$L^t(z_t) = \frac{1}{2}||z_t - y_t||_2^2$$

$$= \frac{1}{2}(z_t - y_t)^T(z_t - y_t) \tag{3.39}$$

All other terms have the tuning (or hyper-)parameters β_x, β_h, γ_1 and γ_2 as scaling penalty factors between 0 and 1 to emphasize the importance one places on these individual component losses (or costs). The second two loss terms are, respectively, on the internal state and its hidden activation function (usually, one needs to use only one or the other). One may use either one with an **unsupervised loss function**, e.g., to optimize the **entropy**, **cross-entropy**, or **sparsity** of this internal (state) representation. Here, we set β_h to zero and choose the component loss function on the internal state to be the entropy defined as

$$L^t(x_t) = -\ln|p_{x_t}(x_t)|$$

$$\approx ||x_t||_1 \tag{3.40}$$

where one may use the L_1 norm to approximate the entropy (i.e., one imposes the **sup-Gaussian Laplacian density function** on the state vector x_t). Another candidate is to use the general approximating function for sup-Gaussian densities, namely, $tanh(\alpha * x_t)$, $1 < \alpha \le 3$, for further details, see, e.g., Waheed and Salem (2003) and Albataineh and Salem (2016).

Finally, the loss function terms on the parameters θ_t and v_t can be used for **regularization**. In that case, one provides a scaled quadratic expression for every scalar parameter. A common example is to use

$$L^t(\theta_t) = \frac{1}{2}||\theta_t||_2^2, \quad 0 \le t \le N-1$$

and for the parameters in the output layer,

$$L^t(v_t) = \frac{1}{2}||v_t||_2^2, \quad 0 \le t \le N$$

These specific loss function terms represent common choices in neural networks generally, including **regularization** in RNNs.

3.4   Basic Recurrent Neural Networks (bRNN): Summary Equations

We now summarize the set of equations to be used in computations as follows:
The basic RNN (bRNN) architecture is

$$x_{t+1} = Ax_t + U_k h_t + W_k s_t + b_k, \quad x_o, \quad t = 0, \cdots, N-1$$

$$h_t = \sigma_t(x_t), \qquad\qquad\qquad\qquad t = 0, \cdots, N \qquad (3.41)$$

$$z_t = V_k h_t + D_k s_t + c_k, \qquad\qquad\quad t = 0, \cdots, N$$

where the adaptive parameters are now indexed with the **iteration** (e.g., **mini-batch**) index k to denote when these parameters are updated during training. These parameters are expected to converge to a near constant at the end of the training process (e.g., a number of **epochs**). The **iteration** index k denotes the mini-batch or the number of samples, where each sample is a full sequence over the finite-horizon training duration. Specifically, a sample is composed of the input-label pairs (s_t, y_t) over the finite-time horizon $0 \le t \le N$.

We thus use the time index t along each (time) step of the sample sequence over the finite horizon, $0 \le t \le N$, while the parameter-update **iteration** index k per one, or multiple, full sample sequence as illustrated in the figure below:

$$- | - - - - - - - - - - - - - - - \rightarrow t \quad \Leftarrow \{\text{sample index}\}$$

$$|$$

$$|$$

$$\{\text{iteration index}\} \Rightarrow k \downarrow$$

For a fixed **iteration** index k, this RNN propagates forward for a given (or chosen) initial condition x_o, and a sequence pair (s_t, y_t) to generate the state x_t, the corresponding hidden nonlinearity h_t, and the output z_t over the sequence duration $t = 0, \ldots, N$.

The output error sequence $z_t - y_t$ is then determined. The backpropagation co-state equations need the final co-state and the output error sequence to generate the co-states λ_t backward in time t over $N \ge t \ge 1$. Here, the final co-state is expressed as (see 3.37)

$$\lambda_N = (\sigma_N')^T V_k^T (z_N - y_N)$$

where σ_N' is the derivative of the nonlinear vector σ_N that is represented as a diagonal matrix of element-wise derivatives of the corresponding scalar nonlinearity. It may also be represented as a vector of derivatives that point-wise multiplies the transformed error vector as a vector–vector **Hadamard multiplication**. Usually, in codes, the latter representation is followed. However, for analysis, it is more convenient to view it as a diagonal matrix (transpose) as it follows from the vector calculus. Let us compactly express the output error sequence over the (finite-)time horizon duration, i.e.,

$$e_t = (z_t - y_t), \quad 0 \leq t \leq N \tag{3.42}$$

Then the final-time boundary condition of the co-state is expressed as

$$\lambda_N = (\sigma'_N)^T V_k^T e_N \tag{3.43}$$

Now the backpropagating co-state dynamics are (see Appendix 3.2)

$$\lambda_t = (A + U_k \sigma'_t)^T \lambda_{t+1} + (\sigma'_t)^T V_k^T e_t +$$
$$\beta_x \left(\frac{\partial L^t}{\partial x_t} \right) + \beta_h (\sigma'_t)^T \left(\frac{\partial L^t}{\partial h_t} \right),$$
$$0 < t \leq N - 1 \tag{3.44}$$

From Appendix 3.2 at the end of this chapter, the state equation parameter gradient changes *at each time instance t*, $0 \leq t \leq N - 1$ (without regularization) are

$$\Delta U_t = -\eta \lambda_{t+1} (h_t)^T \tag{3.45}$$

$$\Delta W_t = -\eta \lambda_{t+1} (s_t)^T \tag{3.46}$$

$$\Delta b_t = -\eta \lambda_{t+1} \tag{3.47}$$

Similarly, from Appendix 3.2, the parameter (weight and bias) updates in the output equation *at each time instance t*, $0 \leq t \leq N$ (without regularization) are

$$\Delta V_t = -\eta \left(\frac{\partial L^t}{\partial V_t} \right) = -\eta e_t (h_t)^T \tag{3.48}$$

$$\Delta D_t = -\eta \left(\frac{\partial L^t}{\partial D_t} \right) = -\eta e_t (s_t)^T \tag{3.49}$$

$$\Delta c_t = -\eta \left(\frac{\partial L^t}{\partial c_t} \right) = -\eta e_t \tag{3.50}$$

Usually, these parameter-change contributions at each t instance are summed over t over their time horizon to provide one change per sample (i.e., a sequence over the time horizon duration). Then, one may update all the parameters after one sample, known as **online learning**, or multiple samples, known as a **mini-batch**, to update the **iteration** index k in the training. Once the iteration parameter updates converge to near constant values, or iterations reach a pre-set number of **mini-batches** or **epochs**, then training is stopped and the parameters are *frozen* in order to provide a finalized **bRNN model** with fixed parameters, ready for testing or **inference**.

Thus, e.g., per sample, i.e., a sequence over the time horizon, one can accumulate the parameter-change contributions to obtain their updates as

$$\Delta U_k = -\eta \sum_{t=o}^{N-1} \lambda_{t+1}(h_t)^T \tag{3.51}$$

$$\Delta W_k = -\eta \sum_{t=o}^{N-1} \lambda_{t+1}(s_t)^T \tag{3.52}$$

$$\Delta b_k = -\eta \sum_{t=o}^{N-1} \lambda_{t+1} \tag{3.53}$$

Similarly, the parameter (weight and bias) updates in the output equation at one iteration (in this case a single sample sequence) are

$$\Delta V_k = -\eta \sum_{t=o}^{N} \left(\frac{\partial L^t}{\partial V_t}\right) = -\eta \sum_{t=o}^{N} e_t(h_t)^T \tag{3.54}$$

$$\Delta D_k = -\eta \sum_{t=o}^{N} \left(\frac{\partial L^t}{\partial D_t}\right) = -\eta \sum_{t=o}^{N} e_t(s_t)^T \tag{3.55}$$

$$\Delta c_k = -\eta \sum_{t=o}^{N} \left(\frac{\partial L^t}{\partial c_t}\right) = -\eta \sum_{t=o}^{N} e_t \tag{3.56}$$

Remark 3.7 The above parameter-update expressions are the sum of the contributions over the time index t, $0 \le t \le N$, i.e., over the full sequence, which is proportional to the time-averaged value. We note that this is only one choice. Observe also that it is equivalent to using the *mean*, or more precisely the time average, of the gradient changes over the sequence of Eqs. (3.45)–(3.50). Note that the hyper-parameter η absorbs any scaling due to the number of elements of the sequence. The mean, however, may be small or even zero, while some gradient changes at some time indices t may be relatively large (in absolute value). In fact, this is one explanation of the "**vanishing gradient**" phenomena. That is, even though the **co-state** and output error sequences are not vanishing, the gradient changes using the "mean" as in Eqs. (3.51)–(3.56) are.

Remark 3.8 Other choices to explore are to use the *median*, the *minimum*, or the *maximum* of the changes over the gradient change sequences, Eqs. (3.45)–(3.50).

Moreover, the change in power or strength in the sequence of changes is best captured by the variance, which should influence the learning rate in a multitude of ways. Thus, there are alternate possibilities of the iteration update laws for RNN besides the ones in Eqs. (3.51)–(3.56).

Remark 3.9 The "**vanishing and exploding**" gradients can be explained easily by investigating the update equations (3.51)–(3.56), and the backward co-state dynamics equations (3.43)–(3.44). We shall of course assume that the network has a sufficiently large dimension with the capacity to *learning* the desired **sequence-to-sequence** (S-2-S) mapping. This is an existence property, i.e., there exist sets of parameters that enable the S-2-S mapping for the network. By their nature, the state and co-state equations form an unstable saddle point dynamics, i.e., if the forward network dynamics are (locally) stable, the backward sensitivity dynamics are also stable but only backward in time. This is another justification for the error backpropagation in addition to the fact that we use the final co-state value to trigger the co-state sequence backward in time. In contrast, if the forward network dynamics are (locally) unstable, the backward sensitivity dynamics are also unstable backward in time. In that event, the state will grow as it propagates forward, thus, the final co-state could become relatively large, and in turn the co-state will be further growing as it propagates backward in time. As the co-state and the error signals are used in the expression of the gradient updates, the gradient will continue to relatively grow unbounded. This process will make the state and co-state grow unbounded. As the time horizon increases (i.e., N increases), this will amplify the growth of the state and co-state leading to the so-called **exploding gradient** phenomena.

Remark 3.10 The leading negative sign in the update laws equations (3.45)–(3.50) can be fully absorbed by redefining both the variables e_t and λ_t to be the negative of what they were defined above. We leave it as an exercise to the reader to re-write the resulting update equations and their consequences.

Exercise Redefine the variable e_t as the negative of the definition above, i.e., redefine it as $e_t = (y_t - z_t), \ 0 \le t \le N$. Show the changes to all subsequent equations in λ_t and the consequent removal of the leading negative sign from the parameter-update laws equations (3.45)–(3.50). (Hint: refine the co-state also by its negative.)

Remark 3.11 In the update laws and the iterations, one must use sufficiently small η to avoid the occurrence of numerical instability—near convergence—while allowing the iteration process of the parameter update to be sufficiently small. The presence of the dominant stable linear (affine) term in bRNN would ensure that, for bounded parameters, the network is **bounded-input-bounded-state** (BIBS) stable and thus the states, co-states, and consequently, the gradient would not grow unbounded. The bRRN can steer the error and the co-state sequences toward zero during training. This is possible of course for all co-states except for the co-state λ_0 that can be non-zero and does not play any role in the (state equations) *parameter updates*, Eq. (3.51)–(3.53).

Exercise   Determine the update laws for the sRNN from the update derivations of the bRNN. (Hint: Set the stable matrix to zero and *undo* the time shift applied to the original sRNN architecture.)

3.5   Concluding Remarks

This chapter introduces the simple recurrent neural networks (sRNN) and identifies the internal state and the hidden (activation) function. It also shows the simple calculations of the **backpropagation through time** (BPTT) for the parameter update during training for supervised learning. However, BPTT does not teach, e.g., how to *incorporate unsupervised loss on the internal state(s)*.

   Then, the chapter follows up with an improved design for a basic recurrent neural network (bRNN) that slightly adjusts the simple RNN (sRNN) in order to improve the dynamic behavior. This entails a basic recurrent structure and an output layer with sufficient parameters to enable a flexible (bandpass) filtering suitable for general RNN applications. The framework adopts the classical constrained optimization and calculus of variations to derive the *generalized* **backpropagation through time** (BPTT) and (stochastic) gradient parameter-update laws. This principled approach enables ease in incorporating general loss functions. The presentation clearly shows the correspondence of the backpropagation through time (BPTT) approach with applying the classical Lagrange multiplier method in constrained optimization. One may choose a loss function as a sum of differing (supervised or unsupervised) loss components on any variable in the recurrent networks, namely outputs, states, or hidden units (i.e., nonlinear functions of the states), and also explicitly on the parameters. It further shows that the usual sum of contributions of changes to the parameters inherent in the BPTT approach is only one form of update, and it could be a source of the "vanishing gradient" phenomena. Assuming a network has the capacity to learn the sequence-to-sequence mapping, the "exploding gradient" phenomena is explained to emanate from either (i) numerical instability due to summing contributions over a long time horizon or (ii) due to the instability of the networks state and the co-state dynamics manifesting themselves in the gradient update laws.

3.6   Appendix 3.1: Global Stability of bRNN

We now show that the presence of the linear (i.e., affine) term with a constant stable matrix in the **basic recurrent neural network** (bRNN) architecture, expressed in Eq. (3.16), is (i) crucial for the boundedness of all trajectories and (ii) all trajectories would remain within, or converge to, a bounded region. This guarantees that, for bounded-input signals and bounded parameters, (the internal state) trajectories would not go unbounded, but rather remain confined to, or converge to, a bounded region.

The bRNN stability and dynamic behavior are governed by the dynamic equation (3.16). To pursue the stability analysis, we define the (bounded) vector:

$$\bar{m}_t := U h_t + W s_t + b \tag{3.57}$$

All components of this vector $\bar{m}_t$ are bounded by *assumption*. We also define the constant matrix A to be a (discretely) *stable* matrix, i.e., it has eigenvalues having moduli less than unity (in absolute value). For instance, $A = \alpha I$, where I is the identity matrix and $0 < \alpha < 1$ is a simple choice.

Then, Eq. (3.16) is now re-written as

$$x_{t+1} = A x_t + \bar{m}_t, \qquad\qquad t = 0, \cdots, N - 1 \tag{3.58}$$

For stability investigation of this now linear system (Antsaklis & Michel, 2006), we choose the following general quadratic form as a candidate **Liapunov function**:

$$E_t = (x_t - x_1^*)^T S (x_t - x_1^*) \tag{3.59}$$

where S is any **symmetric positive definite matrix**, and x_1^* is a center point that may be different from the origin. In Liapunov theory, typically, x_1^* is chosen to be an equilibrium point of interest. In the present case, however, x_1^* is simply a point in (the state) space as a center to be determined. We note that this Liapunov candidate is positive definite for all points with reference to the center point x_1^*.

Now, we calculate the "**difference equation**" of this Liapunov function, i.e.,

$$\begin{aligned} \Delta E_t &= E_{t+1} - E_t \\ &= (x_{t+1} - x_1^*)^T S(x_{t+1} - x_1^*) - \\ &\quad (x_t - x_1^*)^T S(x_t - x_1^*) \end{aligned} \tag{3.60}$$

We use the dynamic equations, Eq. (3.58), into the equality Eq. (3.60) and expand terms to obtain a quadratic equation in x_t. We then "complete the square" and seek to reformulate the expression into the following general quadratic form:

$$\Delta E_t = - \left(x_t - x_2^* \right)^T G^T S G \left(x_t - x_2^* \right) + \bar{d} \tag{3.61}$$

where the **square matrix** G, the vector x_2^*, and the scalar $\bar{d}$ are to be determined in terms of the known quantities in Eq. (3.60). A simple approach in determining these unknowns is to match the expanded equation terms, term by term, as relegated to the following exercise.

Exercise Show that by expanding the expression on the right-hand side of Eq. (3.60), expanding the right-hand side of Eq. (3.61), and then equating corresponding terms in x_t, one obtains the equalities:

$$G^T SG = S - A^T SA \tag{3.62}$$

$$(G^T SG)x_2^* = A^T S(\bar{m}_t - x_1^*) + Sx_1^* \tag{3.63}$$

$$\bar{d} = (\bar{m}_t - x_1^*)^T S(\bar{m}_t - x_1^*) + \{x_2^{*T}(G^T SG)x_2^* - x_1^{*T} Sx_1^*\} \tag{3.64}$$

Hence, for a given or chosen value of the **symmetric matrix** S, and the center vector value x_1^*, the above equalities can provide a solution for a square matrix G, then the center vector x_2^*, and then the scalar $\bar{d}$ used in the **difference Liapunov function**.

Hence, the difference Liapunov function will become **negative definite** outside the **ellipsoid** defined by Eq. (3.61) with the left-hand side set to zero. Let us call this ellipsoid, Ellipsoid B. In graphical form, the ellipsoid centered about x_1^* as defined by the Liapunov function in Eq. (3.59) must expand to engulf Ellipsoid B. Then, at that point, one has another ellipsoid centered around the point x_1^*, let us call it Ellipsoid A, which engulfs Ellipsoid B. Thus, outside of Ellipsoid A, defined by the positive definite Liapunov function, the difference Liapunov function is strictly **negative definite** . Thus, by **Liapunov theory** , all trajectories of this neural network for "frozen" parameters must remain in, or converge to, Ellipsoid A. Thus, global stability (and convergence to Ellipsoid A) is established.

One needs to solve the set of equalities (3.62)–(3.64). To be specific, and to simplify the calculations, we choose $S = I$, the **identity matrix**, the vector $x_1^* = 0$, and proceed as follows. Equation (3.62) becomes

$$G^T G = I - A^T A \tag{3.65}$$

Observe that the square matrix A is "stable," by choice, i.e., it has moduli less than 1. By the singular value decomposition, one obtains

$$A^T A = Q\Gamma^T \Gamma Q^T \tag{3.66}$$

where Γ is a diagonal matrix of the square roots of the singular values of A, and Q is the matrix formed of the corresponding orthogonal right-singular vectors. Thus

$$G^T G = I - A^T A = Q^T [I - \Gamma^T \Gamma]Q > 0 \tag{3.67}$$

which provides a (symmetric) positive definite matrix.

Then to match Eqs. (3.60)–(3.61), term by term, the following equalities must be satisfied: From Eqs. (3.63) and (3.64) in the above exercise, one obtains

$$G^T G = I - A^T A > 0 \tag{3.68}$$

$$(G^T G)x_2^* = A^T \bar{m}_t \tag{3.69}$$

$$\bar{d} = \bar{m}_t^T \bar{m}_t + (x_2^*)^T A^T \bar{m}_t \tag{3.70}$$

The above results are summarized in the following final solution equalities

$$G^T G = I - A^T A > 0 \tag{3.71}$$

$$x_2^* = (G^T G)^{-1} A^T \bar{m}_t \tag{3.72}$$

$$\bar{d} = \bar{m}_t^T \bar{m}_t + (x_2^*)^T A^T \bar{m}_t \tag{3.73}$$

Thus all terms are well-defined in the difference equation (3.61). Hence, we indeed found a Liapunov function candidate centered around $x_1^* = 0$, and it is described over a spherical (or more generally, an ellipsoidal) region. Such region can be made to conservatively include the Ellipsoid B around the point x_2^* as defined above, where outside the bounded region defined by Ellipsoid A, the difference Liapunov equation along all the trajectories is **negative definite**. This ensures that all trajectories of the network will (remain in or) converge to Ellipsoid A thus establishing global stability and convergence.

We remark that the preceding discussion here is to simply determine a bounded region where global stability is established. In general, and for a tighter bound constraints, the goal would be to solve the general constraints in order to determine a point x_1^* as near as possible, or equal to, the point x_2^* where the constraints are satisfied.

Exercise   In order to minimize the size of Ellipsoid A, its center needs to be as close as possible, or equal to, the center of Ellipsoid B, i.e., the vectors x_1^* and x_2^* need to be as close as possible or even equal. Assume $S = I$, where I is the **identity matrix**, and set $x_2^* = x_1^*$, find solutions to Eqs. (3.62–3.64). (Note that while the centers are the same, the ellipsoids would have different shapes/orientations in general).

Finally, we remark that, in the training process of RNNs, the parameters are frozen over the finite-horizon forward state dynamics and backward co-state dynamics. This constitutes one iteration. The gradient descent update of the parameters occurs per one or many iterations (i.e., mini-batch) over the finite-horizon training sequences. This constitutes a decoupling of the dynamics of the bRNN and the parameter gradient updates during training. Thus, the global stability of the overall system is maintained.

3.7   Appendix 3.2: Update Laws Derivation Details

We now provide specific details of the architecture and its parameter updates. We assume the state initial condition x_o is chosen (or given) as a constant, and thus $dx_o = 0$. The **split boundary conditions** then are

$$x_o, \quad \lambda_N = \frac{\partial \phi}{\partial x_N} \tag{3.74}$$

Thus, the process is to use the initial state x_o, the input signal sequence s_t, and the corresponding desired sequence y_t to generate, *forward in time*, the internal state and output sequences.

The final-time loss function $\phi(z_N)$ is defined using the L_2 norm squared:

$$\phi(z_N) = \frac{1}{2}\|z_N - y_N\|_2^2$$

$$= \frac{1}{2}e_N^T e_N \tag{3.75}$$

where we defined the output error by $e_N := z_N - y_N$. Thus its derivative in Eq. (3.74) provides the final co-state as

$$\lambda_N = \frac{\partial \phi}{\partial x_N} = \left(\frac{\partial z_N}{\partial x_N}\right)^T \frac{\partial \phi}{\partial z_N} \tag{3.76}$$

$$= (\sigma_N')^T V_N^T e_N \tag{3.77}$$

For simplicity of derivations, the general loss function is expressed as a sum of separate loss function components as

$$L^t(z_t, x_t, h_t, \theta_t, v_t) = L^t(z_t)$$
$$+ \beta_x L^t(x_t) + \beta_h L^t(h_t)$$
$$+ \gamma_1 L^t(\theta_t) + \gamma_2 L^t(v_t) \tag{3.78}$$

where the loss $L^t(z_t)$ is defined by the L_2 norm (squared)

$$L^t(z_t) = \frac{1}{2}\|z_t - y_t\|_2^2$$

$$= \frac{1}{2}e_t^T e_t \tag{3.79}$$

where we again defined the error signal $e_t := (z_t - y_t)$, $0 \le t \le N$. We now calculate the derivatives needed in the co-state equation (3.25) and in the gradient descent parameter-update Eqs. (3.26) and (3.27), specialized to each parameter.

$$\frac{\partial L^t}{\partial x_t} = \frac{\partial L^t(z_t)}{\partial x_t} + \beta_x \frac{\partial L^t(x_t)}{\partial x_t} + \beta_h \frac{\partial L^t(h_t)}{\partial x_t}$$
$$= (\sigma_t')^T V_t^T e_t + \beta_x \left(\frac{\partial L^t}{\partial x_t}\right) + \beta_h (\sigma_t')^T \left(\frac{\partial L^t}{\partial h_t}\right),$$

$$0 < t \le N - 1 \tag{3.80}$$

And the derivative (for the network Jacobian):

$$\left(\frac{\partial f^t}{\partial x_t}\right) = (A + U_t(\sigma_t)'), \quad 0 < t \leq N - 1 \tag{3.81}$$

Also, one derives (e.g., using scalar calculus and matrix properties) the equalities:

$$\frac{\partial L^t}{\partial \theta_t} = \gamma_1 \theta_t \tag{3.82}$$

$$\left(\frac{\partial f^t}{\partial U_t}\right)^T \lambda_{t+1} = \lambda_{t+1}(h_t)^T \tag{3.83}$$

$$\left(\frac{\partial f^t}{\partial W_t}\right)^T \lambda_{t+1} = \lambda_{t+1}(s_t)^T \tag{3.84}$$

$$\left(\frac{\partial f^t}{\partial b_t}\right)^T \lambda_{t+1} = \lambda_{t+1} \tag{3.85}$$

Thus, this gives

$$\Delta U_t = -\eta \left(\gamma_1 U_t + \lambda_{t+1}(h_t)^T\right) \tag{3.86}$$

$$\Delta W_t = -\eta \left(\gamma_1 W_t + \lambda_{t+1}(s_t)^T\right) \tag{3.87}$$

$$\Delta b_t = -\eta \left(\gamma_1 b_t + \lambda_{t+1}\right) \tag{3.88}$$

The parameter (weight and bias) updates in the output equation are now calculated. The regularization terms for all parameters contribute

$$\frac{\partial L^t(v)}{\partial v_t} = \gamma_2 v_t \tag{3.89}$$

Finally, the gradient changes at each time step t for the linear output layer general parameters are

$$\Delta v_t = -\eta \frac{\partial H^t}{\partial v_t} = -\eta \left(\frac{\partial L^t}{\partial v_t}\right) \tag{3.90}$$

which specializes to the specific two matrices and vector as

$$\Delta V_t = -\eta \left(\frac{\partial L^t}{\partial V_t} \right) = -\eta \left(\gamma_2 V_t + e_t (h_t)^T \right) \tag{3.91}$$

$$\Delta D_t = -\eta \left(\frac{\partial L^t}{\partial D_t} \right) = -\eta \left(\gamma_2 D_t + e_t (s_t)^T \right) \tag{3.92}$$

$$\Delta c_t = -\eta \left(\frac{\partial L^t}{\partial c_t} \right) = -\eta \left(\gamma_2 c_t + e_t \right) \tag{3.93}$$

Thus, the training process is to use the same initial state x_o, and the input signal sequence s_t and its corresponding desired sequence y_t to (i) perform the network time-forward pass, and (ii) to use the boundary condition to perform the time-backward pass, and (iii) generate the gradient descent parameter changes. Then one updates the parameter(s) online over one **sample**, i.e., over the full sequence finite-time horizon, or over several samples (i.e., one a **mini-batch**) to steer the parameters along a (approximate) stochastic gradient descent toward an acceptable "good" local minimum.

Chapter 4
Gated RNN: The Long Short-Term Memory (LSTM) RNN

4.1 Introduction and Background

There are three main gating architectures for Recurrent Neural networks (RNN) in "**Deep Learning**," with impressive demonstrated performance in sequence-to-sequence (S-2-S) applications (Bengio et al., 1994; Chung et al., 2014b; Gers et al., 2002; Graves, 2012; Greff et al., 2017; Hochreiter & Schmidhuber, 1997; Johnson et al., 2016; Zaremba, 2015; Zhou et al., 2016). These are known as **LSTM**, **GRU**, and **MGU** RNN. This chapter is dedicated to the (standard) LSTM RNN, the dominant workhorse in sequence processing. LSTM RNN use three *separate* gating signals. Each gating signal is itself a replica of a simple recurrent neural network (sRNN) with its own parameters (at least *two* matrices and a bias vector). Specifically, each gating signal is an output of a logistic nonlinearity driven by a weighted sum of at least *three* terms: (i) one adaptive weight matrix multiplied by the incoming (external) sequence vector, (ii) one adaptive weight matrix multiplied by the previous memory/activation state vector, and (iii) one (adaptive) bias vector. This is the basic composition driving the gating mechanism in the **gated RNN** architectures literature. There are a host of variants starting from **simple RNN** (sRNN), to **basic RNN (bRNN)** (Salem, 2016a), to more complex **gated RNN variants**, see, e.g., Graves (2012), Zaremba (2015), Greff et al. (2017), and Salem (2018).

4.2 The Standard LSTM RNN

For the purpose of a reference LSTM, we will describe the base or **standard LSTM RNN**. Recall that the dimension of the input (sequence) vector s_t is denoted as m, and the dimension of the hidden unit, and similarly the state (or memory), vector is denoted as n.

© The Author(s), under exclusive license to Springer Nature Switzerland AG 2022
F. M. Salem, *Recurrent Neural Networks*,
https://doi.org/10.1007/978-3-030-89929-5_4

We begin with the so-called simple RNN (sRNN), i.e.,

$$h_t = g(U h_{t-1} + W s_t + b) \tag{4.1}$$

where s_t is the (external) m-dimensional input vector at time step t, h_t is the n-dimensional hidden unit or activation state, g is the (point-wise) activation function, such as the logistic function, the hyperbolic tangent function, or the rectified Linear Unit (ReLU), see, e.g., Chung et al. (2014b), Zaremba (2015), and U, W, *and* b are the appropriately sized parameters (namely, two weight matrices and a bias). Specifically, in this case, U is an $n \times n$ matrix, W is an $n \times m$ matrix, and b is an $n \times 1$ matrix (or vector).

Bengio et al. (1994) report that it is difficult to capture **long-term dependencies** using such simple RNN because, with long sequences, the (averaged stochastic) gradients tend to either vanish or explode. The Long Short-Term Memory (LSTM) RNN, see Hochreiter and Schmidhuber (1997), Graves (2012) and Gers et al. (2002), is the first proposed network architecture to mitigate the "**vanishing**" and/or "**exploding**" gradient problems.

4.2.1   The Long Short-Term Memory (LSTM) RNN

The LSTM RNN architecture introduces the "**memory cell**" to augment the simple RNN architecture of Eq. (4.1). Further, it introduces the gating (control) signals to intuitively (and potentially) steer and mitigate challenges to the gradient computations (during training). Let the simple RNN computation produce its contribution to an intermediate variable, say $\tilde{c}_t$, and add it in a weighted sum to the previous value of the **internal memory state**, say c_{t-1}, to produce the current value of the memory-cell (state) c_t. These operations are expressed as the following set of discrete dynamic equations:

$$\tilde{c}_t = g(U_c h_{t-1} + W_c x_t + b_c) \tag{4.2}$$

$$c_t = f_t \odot c_{t-1} + i_t \odot \tilde{c}_t \tag{4.3}$$

$$h_t = o_t \odot g(c_t) \tag{4.4}$$

The weighted sum is implemented in Eq. (4.3) as an element-wise **(Hadamard) multiplication** denoted by $\odot$ to the gating (control) signals i_t and f_t, respectively. The gating signals i_t, f_t *and* o_t denote, respectively, the *input*, *forget*, and *output* gating signals at (discrete) time (real or fictitious), see Hochreiter and Schmidhuber (1997), Graves (2012), and Greff et al. (2017). In Eqs. (4.2) and (4.4), the activation nonlinearity g is typically the hyperbolic tangent function; however, other forms are possible, e.g., the logistic function or the rectified Linear Unit (reLU).

These three control **gating signals** are in fact chosen to be replica of the basic equation (4.1)—with their own replica parameters and simply replacing g by the logistic function. The logistic function limits the gating signals to within 0 and 1. The specific mathematical form of the gating (control) signals is thus expressed as the vector equations:

$$i_t = \sigma(U_i h_{t-1} + W_i x_t + b_i) \tag{4.5}$$

$$f_t = \sigma(U_f h_{t-1} + W_f x_t + b_f) \tag{4.6}$$

$$o_t = \sigma(U_o h_{t-1} + W_o x_t + b_o) \tag{4.7}$$

where σ is the logistic nonlinearity and the subscripted parameters for each gate consist of two matrices and a bias vector. Thus, the total number of parameters (represented as matrices and bias vectors) for the 3 gates and the memory-cell structure are, respectively, W_i, U_i, b_i, W_f, U_f, b_f, W_o, U_o, b_o, W_c, U_c, and b_c. These parameters are all updated at each training iteration step (i.e., a minibatch). It is immediately noted that the number of parameters in the LSTM model is increased 4-folds from the simple RNN model in Eq. (4.1). Let us assume that the cell state c_t is n-dimensional. (Note that the activation and all the gates have the same dimensions). The input signal is assumed to be m-dimensional. Then, the total parameters in the LSTM RNN is equal to $4 \times (n^2 + nm + n) = 4n(n + m + 1)$.

Figure 4.1 provides an illustrative diagram of the memory cell in LSTM with the 3 gates, each represented by a sRNN block.

Remark 4.1 It is worth mentioning that one should choose the state as the memory cell c_t, and thus the activation function h_t is a nonlinear, usually compressive, function of the state. In some LSTM forms both the state and the activation function are incorporated into the gating signals. However, that is technically redundant. One

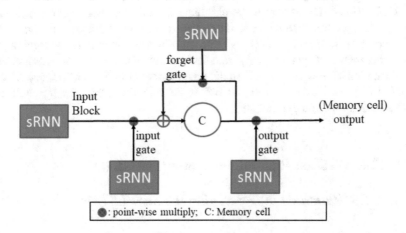

Fig. 4.1   Block-Diagram of the LSTM Memory Cell with the Input Block and 3 Gates

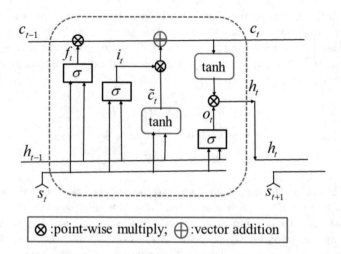

Fig. 4.2 Wiring-Diagram of the LSTM Memory Cell with the Input Block and 3 Gates

may use one or the other. In the equation forms above, labeled as the standard form, one incorporates the (compressed) activation function h_{t-1}; however, c_{t-1} may be used instead. We include an illustrative schematic diagram representation for the LSTM architecture where the "inputs" to the LSTM block are the present (external) input vector s_t, and the previous state c_{t-1} and the activation state h_{t-1} and the "outputs" of this block are the present state c_t and the activation state h_t—see Fig. 4.2.

LSTM RNNs have demonstrated their effectiveness in practice involving sequences in numerous and varied applications. A sample of their effectiveness since their publication in Hochreiter and Schmidhuber (1997) is exhibited in Greff et al. (2017) and the references therein. Indeed, their large scale deployment into the first Google-translate commercial system reflects their practical prowess, see Johnson et al. (2016). However, as large scale neural systems are becoming the norm, the matter of quadrupling parameters has become a relative constraint. For an illustration, if $m = n = 400$, then the total number of parameters per one LSTM RNN is $N = 4(400)(801) = 1,281,600$ to be trained via, e.g., the (generalized) backpropagation through time (BPTT)!

4.3   Slim LSTMs: Reductions within the Gate(s)

4.3.1   The Rationale in Developing the Slim LSTMs

A key point is to recognize and exploit the role of the internal dynamic "state" that captures the essential information about processing an input signal's time history.

As the state contains the essential summary information about a network, including the input sequence time history, one can eliminate (redundant) terms that are already contained in the state, directly or indirectly. Limiting attention to the gating signals, during training, the two weights and bias vector update laws depend on the external input signal and/or the previous memory state(s)—directly or indirectly. Thus, there is redundancy in using all three terms in the gates to generate the gating signal(s). Exploiting this observation allows for the development of several variant networks with reduced parameters resulting in appreciable computational savings.

The view is to consider the gates as control "signals" which essentially only needs a measure (or a function) of the network's state. To achieve effective learning toward a desired (low loss or) high accuracy performance, one only needs (a function of) the state as it captures the history of the input sequence. In that view, the form of the standard LSTM network is overly redundant in generating such control signals. For example, it is redundant to provide the state (which may be represented by the memory cell or the activation unit, but not both!) and the external input signal to the gating signal.

For one, the derived (gradient descent) update learning law(s) of the bias vector itself depends on the co-state vector, which depends on the history of the output and memory state vector (which capture the time history of the input vector). The co-state and state vector pairs, again, capture all information pertaining to the signals in the dynamic system history profile—specifically the external input (time-) sequence. A present input value (in the gating signal) may add a new information to prior state values; however, it may also bring an instantaneous outlier value corrupted by noisy measurements or external noise. On the other hand, one notes that in certain scenarios or applications, the instantaneous or present input value could be valuable for quick reaction to new (or sudden) changes in sensory signals specially in online training or adaptation. For the purposes of reducing parameterization here, we choose forms that eliminate the instantaneous input signal from (all) gating signals. The intent is to strife to retain the accuracy performance of a gated RNN while aggressively reducing the number of (adaptive) parameters to various degrees. Such parameter-reduced architectures would speedup execution in training and **inference modes** and may be more suitable for limited embedded or mobile computing platforms.

From a recurrent dynamic systems view, the qualitative performance is expected to be retained. However, the quantitative performance would of course be varied as the number of parameters is reduced to different degrees.

The gating signals in **gated RNN** enlist all of (i) the previous hidden unit and/or state, (ii) the present input signal, and (iii) a bias, in order to enable the gated RNN to effectively acquire the capability to learn sequence-to-sequence (S-2-S) mappings. The dominant adaptive algorithms used in training are essentially varied forms of backpropagation through time (BPTT) stochastic gradient descent. Each gate simply replicates a simple RNN. All parameters in this LSTM structure are updated using the overall network BPTT stochastic gradient descent to minimize a loss function, see, e.g., Gers et al. (2002), Greff et al. (2017). The concept of state, which in essence summarizes the information of the gated RNN up to the present (or in other manifestations, the previous) time step, contains the information about the time history of the input sequence. Moreover, a parameter update also includes information pertaining to the co-state (and state) of the overall network structure, see chapter 3 and also Salem (2016a, 2016b).

For tractable and **modular realizations**, we apply the modifications to all gating signals uniformly, i.e., the same modification(s) is applied to all gates.

A gating signal is driven by 3 components, resulting in 8 possible variations—including the trivial one when all three components are absent. Without the external input signal, there are 3 non-trivial variants per gate. For efficiency, we consider the 3 variants without the external input sequence as the input sequence over its time horizon is captured within the "state."

4.3.2   Variant 1: The LSTM_1 RNN

In this variant, each gate is computed using the previous hidden activation (as the nonlinear function of the state) and the bias, thus reducing the total number of parameters from the 3 gate signals, in comparison to the LSTM RNN, by $3 \times nm$.

$$i_t = \sigma(U_i h_{t-1} + b_i) \qquad (4.8)$$

$$f_t = \sigma(U_f h_{t-1} + b_f) \qquad (4.9)$$

$$o_t = \sigma(U_o h_{t-1} + b_o) \qquad (4.10)$$

4.3.3   Variant 2: The LSTM_2 RNN

In this variant, each gate is computed using only the previous hidden state, thus reducing the total number of parameters from the 3 gate signals, in comparison to the LSTM RNN, by $3 \times (nm + n)$.

$$i_t = \sigma(U_i h_{t-1}) \qquad (4.11)$$

$$f_t = \sigma(U_f h_{t-1}) \qquad (4.12)$$

$$o_t = \sigma(U_o h_{t-1}) \qquad (4.13)$$

4.3.4   Variant 3: The LSTM_3 RNN

In this variant, each gate is computed using only the bias, thus reducing the total number of parameters in the 3 gate signals, in comparison to the LSTM RNN, by $3 \times (nm + n^2)$.

$$i_t = \sigma(b_i) \qquad (4.14)$$

$$f_t = \sigma(b_f) \qquad (4.15)$$

$$o_t = \sigma(b_o) \qquad (4.16)$$

In order to reduce the parameters even further, one replaces the standard multiplication by point-wise (Hadamard) multiplication. In the case of the hidden units, the matrices U_* are reduced into (column) vectors of the same dimension as the hidden units (i.e., n). We denote these corresponding vectors by u_* as delineated next.

4.3.5   Variant 4: The LSTM_4 RNN

In this variant, each gate is computed using only the previous hidden state but with point-wise multiplication. Thus one reduces the total number of parameters, in comparison to the LSTM RNN, by $3 \times (nm + n^2)$.

$$i_t = \sigma(u_i \odot h_{t-1}) \qquad (4.17)$$

$$f_t = \sigma(u_f \odot h_{t-1}) \qquad (4.18)$$

$$o_t = \sigma(u_o \odot h_{t-1}) \qquad (4.19)$$

4.3.6   Variant 5: The LSTM_5 RNN

In this variant, each gate is computed using only the bias plus the previous hidden state with point-wise multiplication as follows:

$$i_t = \sigma(u_i \odot h_{t-1} + b_i) \qquad (4.20)$$

$$f_t = \sigma(u_f \odot h_{t-1} + b_f) \qquad (4.21)$$

$$o_t = \sigma(u_o \odot h_{t-1} + b_o) \qquad (4.22)$$

We remark that in the above variants as well as in the standard LSTM RNN, one can replace the bounded activation state h_* by the cell state c_* with similar qualitative behavior. However, the specific quantitative behavior is expected to differ.

For more forms of *slim* LSTM RNN, we refer the reader to Salem (2016b, 2018). In practice, when considering an application, it is prudent to begin with the *standard* LSTM RNN or other gated RNN available on your "deep learning" **computational framework**. Then, when optimizing for computational load or speed, one may explore other parameter-reduced forms of *slim* LSTM RNN.

4.4   Comparative Experiments of LSTM RNN Variants

We explore sample case-study experiments on the standard LSTM RNNs and the slim LSTM variants for the purpose of initial comparative base performance. The effectiveness of the standard and the slim LSTM RNN has been evaluated on multiple **public datasets** considered to be the "hello-world" of datasets in **neural networks and deep learning**. These are the MNIST and IMDB datasets. The intent here is to demonstrate the performance of the standard LSTM RNN and, as well, the comparative performance of the slim LSTM variants, rather than to achieve **state-of-the-art** results. What is called here the standard LSTM RNN was used as a baseline architecture and compared with its five *slim* variants. For more details and variants, see Greff et al. (2017, 2017a, 2017b, 2017c), Lu and Salem (2017), Salem (2018).

4.4.1   Experiments on the MNIST Dataset

This dataset contains 50,000 training set, 10,000 evaluation set, and 10,000 testing set of handwritten images of the digits (0–9). The training and evaluation sets contain the corresponding labeled class for each digit image. Each (digit) image has a size of 28×28 pixels (or usually 32×32 pixels padded by 2 rows/columns of zeros around the image in anticipation of smoother convolution operations). The (digit) image data were pre-processed to have zero mean and unit variance. As in the work of Le et al. (2015), Jianxin et al. (2016), and also in Chollet (b,c), the dataset was organized in two forms to be the input to an LSTM-based network. The first was to reshape each image as a one dimensional vector with pixels scanned row by row, from the top left corner to the bottom right corner. This results in a relatively long sequence input of length 784 pixel elements. The second form requires no reshaping but treats each row of an image as a vector input, thus giving a much shorter input sequence of length 28 vector elements. The two forms of data re-organization were referred to as *pixel-wise* and *row-wise* sequence inputs, respectively. It is noted that the longer *pixel-wise* sequence is more time consuming in training. In the two training tasks, 100 hidden units and 100 training epochs were used for the pixel-wise sequencing input, while 50 hidden units and 200 training epochs were used for the row-wise sequencing input. Other network settings were kept the same throughout, including the *batch size* set to 32, *RMSprop* optimizer, *cross-entropy* loss, dynamic learning rate (η_t), and early stopping strategies. In particular, for the dynamic learning rate, it was set to be an *exponential* function of the *training* loss to speed up training. Specifically, the (dynamic) learning rate η_t is replaced by the quantity $\eta_o exp(Loss)$, where η_o is a constant coefficient, chosen judiciously, and *Loss* is the training loss, e.g., changed every new mini-batch or every new epoch. The first time the exponential-loss learning rate was introduced, to the best of our knowledge, is in Ahmad and Salem (1992).

For the *pixel-wise* sequence, as it takes relatively longer time to train, only two **learning rate coefficients** $\eta_o = 1e^{-3}$ and $\eta_o = 1e^{-4}$ were considered in this experiment. In contrast, for the *row-wise* sequence, four η_o values of $1e^{-2}, 1e^{-3}, 1e^{-4}$, and $1e^{-5}$ were considered. The **dynamic learning rate** η_t is thus directly (proportional) to the training loss performance. At the initial stage, as the initial parameters were randomly chosen, the training loss is typically large. This typically results in a large effective learning rate (η_t), which in turn increases the stepping of the gradient further from the present parameter (vector) position. The **effective learning rate** η_t decreases only as the (training) loss function decreases toward a lower loss level and eventually toward an acceptable (local) minimum in the parameter space. Thus, it was found to achieve faster convergence to an acceptable solution as anticipated. For the **early stopping criterion**, the training process would be terminated if there was no improvement (or when performance deteriorates) on the evaluation data over consecutive epochs—in these experiments we chose 25 consecutive epochs. We relegate further details of the initiation MNIST experiments to, e.g., Lu and Salem (2017), Akandeh and Salem (2017a), (2017b), (2017c).

4.4.2   Experiments on the IMDB Dataset

This public dataset is the classic **IMDB dataset** acquired here via the Keras framework (Chollet, a). The IMDB Dataset is a binary sentiment classification dataset using 5000 samples for training and 5000 samples for evaluation or testing. In these experiments, the coding was executed using **Keras** with the Tensorflow framework as a backend, see Chollet (a).

Each sample (i.e., review) is truncated or padded to 500 words. The first layer is a (trained) embedding layer which is in fact a simple multiplication or a matrix look-up that maps each word's integer rank identifier to its corresponding word embedding vector representation. For each sample (i.e., review) the embedding will output corresponding vector sequence of length 500 representing the review. The embedding output feeds into the subsequent LSTM layer of a chosen hidden units dimension followed by a dense layer that outputs the confidence of the review belonging to one of two classes (say the positive class). For simplicity, if the confidence is above 0.5, the review is positive, if the confidence is below 0.5, the review is negative. Of course more refined criterion to classify the reviews can be

Table 4.1 Network specifications for the IMDB Datasets

Input dimension	32
Sequence length	500
Number of hidden units	200
Output dimension	1
Nonlinear function	Sigmoid
Loss function	Binary cross entropy
Optimizer	Adam
Batch size	32
Number of epochs	100

Fig. 4.3 Simple Network Architecture Diagram for the IMDB dataset

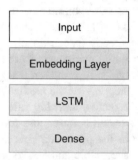

prescribed. For example, one may create an uncertain interval, say [0.4, 0.6], where the output sentiment is uncertain.

The adopted network specification from **Keras** is delineated in Table 4.1. It is followed by a depiction of the (stacked) schematic representation of the barebone network architecture in Fig. 4.3. The network is very simple in order to exhibit the main role of the LSTM RNNs with minimal additional layers. It is composed of the sequential layers of (i) Embedding layer, (ii) LSTM Layer, and (iii) Dense layer to output the confidence of belonging to the class of positive reviews. The input layer takes in the integer rank of the words in a truncated (or padded) sample review up to 500 words and outputs the embedding vectors of dimension 32 in a 500-element sequence. The LSTM takes in the sample (review) sequence representation and outputs one vector of $200 - d$ (dimension), representing the hidden units at the final (time-) instance 500. This is the default position in the LSTM layer in **Keras** where only the last element of the output sequence is returned. This final $200 - d$ vector is subsequently fed to a Dense layer with a single output with sigmoidal nonlinearity. In the training for this experiment, we use the **binary cross-entropy** loss function, the **optimizer Adam**, a batch size of 32, and 100 epochs.

We show plots in Fig. 4.4 resulting from sample experiments of the baseline or *standard* LSTM RNN (denoted in the figure curves as lstm0) vis-a-vis the five *slim* LSTM variants. The hyper-parameters (specially, the learning rate η) were set to be best matched for the *standard* LSTM RNN as suggested from the example codes on **Keras** at the time of the experiments. In Fig. 4.4, one observes that the Training Accuracy curve of lstm0 is higher than all other Training Accuracy curves of the 5 variants. However, lstm0, lstm1, and lstm2 Testing Accuracy

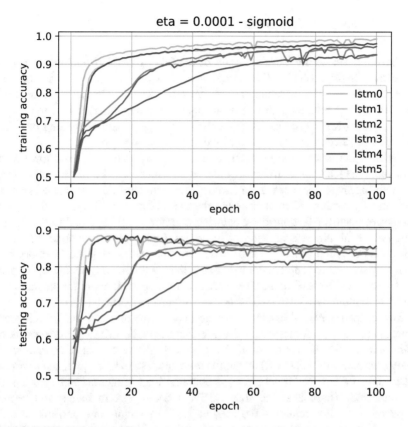

Fig. 4.4 IMDB, Training and Test accuracy, $\sigma = sigmoid$, $\eta = 1e-4$

curves are all practically the same. Then, the remaining Testing Accuracy curves of lstm3, lstm4, and lstm5 are a bit behind. These types of plots are typical in numerous runs. It is noticeable that the performance of the *slim* LSTM variants is very competitive, and some are close to the *standard* LSTM RNN in these experiments. This establishes the viability of using the *slim* LSTM variants specially for limited compute resources of edge devices. More elaborate *slim* LSTM variants and comparative computational case studies can be found in the open literature (Salem, 2018) and the references therein. It should be noted that an elaborate comparison would be to seek to optimize the hyper-parameters (e.g., the learning rate η) for the LSTM RNN and the family of variants. This fine-tuning of each LSTM RNN variant may be a basis for project or deployment challenges.

4.5   Concluding Remarks

We have introduced the basic structure of the **LSTM RNN**. We labeled a version to be the *standard* LSTM RNN as referenced in the literature to set a baseline for comparison. Other baseline choices are possible. One similar baseline would be obtained by replacing the activation state function h_* in the gates by the corresponding (cell) state c_*. This has the immediate benefit of feeding back into the gates the uncompressed version of the network state vector.

We then introduced five reduced-parameter gating signals to strife toward less computational load during training (and also inference). The key observation is that one only needs to feedback a function of the state, and not the input, since the state already carries information about the input sequence. Moreover, there is no qualitative need for feeding back both the (memory-cell) state and the activation state (which in turn is a function of the state). Either one will suffice. While we chose the activation function for feedback into the gates, using the (memory-cell) state may serve some application better as it sheds a nonlinearity in the signal path. There are other forms of *slim* LSTM variants in the open literature for the interested reader to explore, see Salem (2018) and the references therein.

The chapter ends with example experiments to showcase the (training and testing) accuracy performance on the now classic public IMDB dataset for movie reviews to predict the sentiment of the reviewers. The performance comparison is among the *standard* LSTM RNN architecture and five reduced-parameter variants called *slim* LSTM RNNs where computational load reduction and speedup are demonstrated. These example slim LSTM variants seek to reduce redundancy of parameters while retaining the original LSTM structure and performance. As recurrent neural networks are employed to learn sequence-to-sequence (S-2-S) mappings, the question turns to *capacity*, i.e., the existence of a set of parameters in a given architecture (or variant) that enables approximate mappings of finite S-2-S using the training dataset—while generalizing on validation/test data. In that context, the dimension of the LSTM variant would become an important (hyper-) parameter for some *slim* LSTM RNNs that could be used to increase the capacity for improved performance.

Many of these variants have already been validated to produce comparable performance to the standard LSTM RNN in several publications, e.g., (Lu and Salem, 2017; Akandeh and Salem, 2017a, 2017b, 2017c; Salem, 2016b).

Part IV
Gated Recurrent Neural Networks:
The GRU and The MGU RNN

Chapter 5
Gated RNN: The Gated Recurrent Unit (GRU) RNN

5.1 Introduction and Background

Gated Recurrent Neural Networks (Gated RNNs) have shown success in several applications involving sequential or temporal data (Chung et al., 2014b; Zaremba, 2015). For example, they have been applied extensively in speech recognition, music synthesis, natural language processing, machine translation, medical and bio-medical applications, etc., Chung et al. (2014b), Boulanger-Lewandowski et al. (2012). **Long Short-Term Memory** (LSTM) RNNs and the subsequently intro-duced **Gated Recurrent Unit (GRU) RNNs** have been successfully shown to per-form comparatively well with long sequence applications (Boulanger-Lewandowski et al., 2012; Chung et al., 2014b; Gers et al., 2000, 2002; Graves, 2012; Hochreiter and Schmidhuber, 1997; Maas et al., 2011; Mikolov et al., 2014; Zaremba et al., 2014).

Gated RNNs' success is often attributable to the explicit **gate signaling** that implicitly control how the present input-block and previous memory are fused to update the present activation and produce the present state. These **gates** have their own sets of weights that are adaptively updated in the learning phase (i.e., the training and evaluation phases). While these architectures empower successful **learning** in RNNs, they introduce an increase in parameterization through their added **gate-networks**. Consequently, there is an *added* computational expense vis-a-vis the original simple RNN model (Chung et al., 2014b; Hochreiter & Schmidhuber, 1997; Zaremba, 2015). It is noted that the LSTM RNN employs 3 distinct gate-networks while the GRU RNN reduces the gate-networks to two (by simply coupling two gates).

In this chapter, we focus on the **GRU RNN** and provide the original description and terminology as it is initially described in the literature. Then, we explain the GRU RNN in the notation and terminology of the **LSTM RNN**, described in the previous chapter, to show clearly the transition from the LSTM RNN to the GRU RNN. In addition to reducing the gates to two by **coupling** the input-gate and

© The Author(s), under exclusive license to Springer Nature Switzerland AG 2022
F. M. Salem, *Recurrent Neural Networks*,
https://doi.org/10.1007/978-3-030-89929-5_5

the forget-gate by a **direct sum**, the GRU RNN also removes one nonlinearity in the network **signal path** by feeding back the **memory-cell state** instead of the (nonlinear) **activation state**. However, as it was pointed out in the previous chapter, this alternate option of feedback can also be used in the *standard* LSTM RNN.

We subsequently introduce five **gate-variants** for GRU RNN, called *slim* **GRU RNN**, with reduced parameterization and thus reduced computational load. We comparatively evaluate the performance of the *standard* and the *slim* GRU RNNs on two public datasets (Dey & Salem, 2017b). Using the **MNIST dataset** (LeCun et al., 2010), one generates two sequences (Chung et al., 2014b; Hochreiter & Schmidhuber, 1997; Zaremba, 2015; Zhou et al., 2016). One sequence is obtained from each 28×28 image sample as **pixel-wise** long sequence of length $28 \times 28 = 784$ (basically, scanning from the upper left to the bottom right of the image). Also, one generates a **row-wise** short sequence of length 28, with each element being a vector of dimension 28. The third sequence type employs the **IMDB movie review dataset** (Maas et al., 2011) where one chooses the length of the sequence in order to achieve high performance sentiment classification from a given review paragraph.

The remainder of the chapter is organized as follows. Section 5.2 reviews the basic structures of gated RNNs, specifically the popular LSTM and GRU RNNs, and then uses unified notations to elucidate the comparison. Section 5.3 introduces the five variant GRU RNNs, referred to as *slim* **GRU RNNs**. Sect. 5.4 presents and discusses the experiments employing the MNIST and the IMBD datasets to generate time-series of varied lengths (Dey & Salem, 2017b). The comparative results are depicted in performance accuracy curves and summarized in comparative performance tables. Section 5.5 summarizes final concluding remarks.

5.2   The *Standard* GRU RNN

In principle, RNNs are more suitable for capturing relationships among sequential data types. We begin with the so-called simple RNN which has a recurrent hidden state as in

$$h_t = g(U h_{t-1} + W s_t + b) \tag{5.1}$$

where s_t is the (external) m-dimensional input vector at time t, h_t is the $n-$dimensional hidden (activation) state, g is the (element-wise) activation function, such as the logistic function, the hyperbolic tangent function, or the rectified Linear Unit (ReLU) (Chung et al., 2014b; Zaremba, 2015), and U W, *and* b are the appropriately sized parameters (two weights and a bias). In this case, U is an $n \times n$ matrix, W is an $n \times m$ matrix, and b is an $n \times 1$ matrix (or vector).

Hochreiter and Schmidhuber (1997) and Bengio et al. (1994), e.g., argued that it is difficult to capture long-term dependencies using such simple RNNs because the (stochastic) gradients tend to either vanish or explode with long sequences. Two particular architectures: the Long **Short-Term Memory (LSTM) RNNs** (Gers et al., 2002; Hochreiter & Schmidhuber, 1997) and more recently the **Gated Recurrent Unit (GRU) RNNs** (Chung et al., 2014b) have been proposed to solve the "vanishing" or "exploding" gradient problems. While this chapter focuses on the GRU RNN, we will present these two architectures to show the direct relationship and how to (smoothly) transition from the (parent) LSTM RNN to the GRU RNN.

5.2.1 The Long Short-Term Memory (LSTM) RNN

The LSTM RNN architecture uses the computation of the simple RNN of Eq. (5.1) as an intermediate input-block candidate to be added to another internal memory-cell (state). Let us now call the input-block candidate, e.g., $\tilde{c}_t$, and add it in a (element-wise) weighted sum to the previous value of the internal (memory-cell) state, say c_{t-1}, to produce the current value of the (memory-cell) state c_t. This is expressed succinctly in the following **discrete dynamic equations**:

$$\tilde{c}_t = g(U_c h_{t-1} + W_c s_t + b_c) \tag{5.2}$$

$$c_t = f_t \odot c_{t-1} + i_t \odot \tilde{c}_t \tag{5.3}$$

$$h_t = o_t \odot g(c_t) \tag{5.4}$$

In Eqs. (5.2) and (5.4), the activation nonlinearity g is typically the hyperbolic tangent function but may be implemented as a rectified Linear Unit (reLU). The weighted sum is implemented in Eq. (5.3) via element-wise (Hadamard) multiplication denoted by $\odot$ to **gating signals**. The gating (control) signals i_t, f_t, *and* o_t denote, respectively, the *input*, *forget*, and *output* gating signals at time t. These control gating signals are in fact replica of the basic equation (5.2), with their own (distinct) parameters, and replacing g by the logistic function. The logistic function limits the **gating signals** to within 0 and 1 and thus the **gates** intuitively resemble **analog switches**. The specific mathematical form of the gating signals is thus expressed as the vector equations:

$$i_t = \sigma(U_i h_{t-1} + W_i s_t + b_i) \tag{5.5}$$

$$f_t = \sigma(U_f h_{t-1} + W_f s_t + b_f) \tag{5.6}$$

$$o_t = \sigma(U_o h_{t-1} + W_o s_t + b_o) \tag{5.7}$$

where $\sigma(\cdot)$ is the nonlinear logistic function, and the parameters for each gate consist of two matrices and a bias vector. Thus, the total number of parameters (represented as matrices and bias vectors) for the 3 gates and the memory-cell structure are, respectively, U_i, W_i, b_i, U_f, W_f, b_f, U_o, W_o, b_o, U_c, W_c, and b_c. These parameters are all updated at each training step and stored. It is immediately noted that the number of parameters in the LSTM architecture is increased 4-folds from the simple RNN model in Eq. (5.1). (Note that the memory-cell and all the gates have the same dimensions). Assume that the memory-cell state is n-dimensional. Assume also that the input signal is m-dimensional. Then, the total parameters in the LSTM RNN is equal to $4 \times (n^2 + nm + n) = 4n(n + m + 1)$.

5.2.2   The Gated Recurrent Unit (GRU) RNN

The **GRU RNN** reduces the gating signals to two from three in the **LSTM RNN** architecture. The two gates are called an **update gate** z_t and a **reset gate** r_t. The GRU RNN model is first presented in its *original form* (Chung et al., 2014b and the references therein):

$$\tilde{h}_t = g(U_h(r_t \odot h_{t-1}) + W_h s_t + b_h) \tag{5.8}$$

$$h_t = (1 - z_t) \odot h_{t-1} + z_t \odot \tilde{h}_t \tag{5.9}$$

with the two gates presented as:

$$z_t = \sigma(U_z h_{t-1} + W_z s_t + b_z) \tag{5.10}$$

$$r_t = \sigma(U_r h_{t-1} + W_r s_t + b_r) \tag{5.11}$$

One observes that the GRU RNN [Eqs. (5.8)–(5.9)] is similar to the LSTM RNN [Eqs. (5.2)–(5.3)], however with one less external gating signal in the **coupling** interpolation Eq. (5.9). This eliminates one gating signal and the associated parameters. One also notes the presence of only one activation nonlinearity $g(\cdot)$ in the equations (and thus the input signal path).

   In essence, the GRU RNN has a 3-fold increase in parameters in comparison to the simple RNN of Eq. (5.1). Specifically, the total number of parameters in the GRU RNN equals $3 \times (n^2 + nm + n) = 3n(n + m + 1)$. As compared to the LSTM RNN, there is a reduction in parameters of $n(n + m + 1)$. Moreover, it eliminates one (vector) nonlinearity $g(\cdot)$. In various studies, e.g., in Chung et al. (2014b) and the references therein, it has been noted that GRU RNN is comparable to, or may even outperform, the LSTM in some case studies. However, there is no definitive assertion as to in what circumstances these improved performances would arise. That is simply because one cannot. One can only reply on the particular dataset used in the experiments and/or case studies. The main point here is that the GRU RNN has been shown, in specific case studies, to be comparable to the performance of the LSTM RNN. To appreciate the parameter savings in one GRU RNN, if one considers a scenario where $n = m = 100$, then the reduction in parameters from full-fledged LSTM RNN is $N = n(n + m + 1) = 20,100$ less parameters to be adapted and stored.

5.2.3 The Gated Recurrent Unit (GRU) RNN vs. the LSTM RNN

We can adopt the same notation as we used for the LSTM RNNs to facilitate direct comparisons of the architectures. In this view, the LSTM is viewed as the parent architecture from which architectural (as well as parameter) reductions are applied to. This provides a vision of the spectrum of reduction processes that are applied to the most general LSTM RNN. We seek now to express the equations for the GRU RNNs in terms of the common notation of the parent LSTM RNN. Thus, one can easily observe the comparative structure and how one applies reduction to the LSTM RNN to obtain the GRU RNN. To that end, we identify the following notation correspondence among the two gated RNNs:

$$i_t \longleftrightarrow z_t \tag{5.12}$$

$$f_t \longleftrightarrow (1 - z_t) \tag{5.13}$$

$$c_t \longleftrightarrow h_t \tag{5.14}$$

$$\tilde{c}_t \longleftrightarrow \tilde{h}_t \tag{5.15}$$

Thus, using the notation of LSTM RNNs, one can express the GRU RNNs as

$$\tilde{c}_t = g(U_c h_{t-1} + W_c s_t + b_c) \tag{5.16}$$

$$c_t = (1 - i_t) \odot c_{t-1} + i_t \odot \tilde{c}_t \tag{5.17}$$

$$h_{t-1} = o_t \odot c_{t-1} \tag{5.18}$$

where one observes that the gating in the memory-cell is using a single gate, i_t, and its complement $(1-i_t)$. The new activation in Eq. (5.18) has dropped the nonlinearity g in comparison and uses the memory-cell state directly. Thus, the architecture input signal pathway passes through one g nonlinearity in each time cycle. Moreover, the time-step in Eq. (5.18) is not uniform. This one time-step (delay) is not necessarily crucial for the architecture; however, it identifies another distinct difference. It may be revised to be more consistent with the LSTM by using the following activation equation instead—with no anticipation loss in overall performance:

$$h_t = o_t \odot c_t \tag{5.19}$$

In that event, the reduction is streamed directly from the LSTM RNN. Moreover all signals are at the same time-step with potential ease in accessing signal time-samples without the need of storage of previous samples.

The gate signals i_t and o_t are similar to the ones in the LSTM RNN architecture, with a re-focus on the memory-cell state feedback c_t (as opposed to the activation state feedback h_t), specifically,

$$i_t = \sigma(U_i c_{t-1} + W_i s_t + b_i) \tag{5.20}$$

$$o_t = \sigma(U_o c_{t-1} + W_o s_t + b_o) \tag{5.21}$$

From a **nonlinear dynamical systems** viewpoint, these changes in the gate signals will likely produce *quantitatively* different results in simulations under similar conditions; however, they are expected to be *qualitatively* similar in dynamic behavior. The modification to the f_t gate signal may produce *qualitative* behavior in certain scenarios (Salem, 2016a). For ease of comparison among the Gated RNNs, we shall adopt the common notations used here for the LSTM RNNs. However, at times as per convenience and to compare with common results in the literature, we may use the original notations for GRUs, as in Eqs. (5.8–5.11) in this chapter.

This concludes the presentation of the GRU RNN architectures and its direct connection to the parent (in the sense of more inclusive) LSTM RNN architectures.

5.3 Slim GRU RNN: Reductions within the Gate(s)

Analogous to the *slim* LSTM RNNs, we introduce the *slim* GRU RNNs architecture variants. While we are using the original notations for GRU RNNs, we adopt analogous naming as in the *slim* LSTM naming for convenient correspondence.

In this chapter, we focus only on *slim* GRU RNNs with reductions in the gate-networks. For tractable and modular realizations, we consider applying the modifications to the two gating signals i_t and o_t.

Specifically, we retain the architecture of Eqs. (5.8)–(5.9) unchanged and focus on variations in the construct of the gating signals in Eqs. (5.10) and (5.11) or, equivalently, Eqs. (5.20 and 5.21). We apply the variations identically to the two gates for **uniformity** and simplicity.

The gating mechanism in the GRU (and LSTM) RNNs is a replica of the simple RNN in terms of parameterization. The weights corresponding to these gates are also updated using the backpropagation through time (BPTT) stochastic gradient descent as it seeks to minimize a loss function (Gers et al., 2002; Hochreiter & Schmidhuber, 1997). Thus, each parameter update will involve information pertaining to the state of the overall network. Thus, all information regarding the current input and the previous states are reflected in the latest state variable. There is a redundancy in the signals driving the gating signals. The key driving signal should be the internal state of the network. Moreover, the adaptive parameter updates all involve (directly or indirectly) components of the internal state of the system, see Salem (2016b, 2016a). In this section, we consider five distinct variants of the gating equations applied uniformly to both gates. However, non-uniform application to the two gates is also an option.

A gating signal (see, e.g., Eq. (5.20) includes 3 additive terms, which may result in creating 8 possible variations—including the trivial one when all three terms are absent. Without the external input term, there are 3 non-trivial variants per gate. For efficiency, we consider the 3 variants without the external input sample at time t, as the input sequence over its time/sample horizon is captured in the "state." Moreover, an instant input at time t may bring in unwelcome noise or may itself be an outlier.

5.3.1   Variant 1: The GRU_1 RNN

In this variant, each signal gate is computed using the previous (memory-cell) state and the bias, thus reducing the total number of parameters from the two gate signals, in comparison to the GRU RNN, by $2 \times nm$.

$$i_t = \sigma(U_i c_{t-1} + b_i) \tag{5.22}$$

$$o_t = \sigma(U_o c_{t-1} + b_o) \tag{5.23}$$

5.3.2   Variant 2: The GRU_2 RNN

In this variant, each signal gate is computed using only the previous memory-cell state, thus reducing the total number of parameters from the two gate signals, in comparison to the GRU RNN, by $2 \times (nm + n) = 2n(m + 1)$.

$$i_t = \sigma(U_i c_{t-1}) \tag{5.24}$$

$$o_t = \sigma(U_o c_{t-1}) \tag{5.25}$$

5.3.3   Variant 3: The GRU_3 RNN

In this variant, each gate is computed using only the bias, thus reducing the total number of parameters in the two gate signals, in comparison to the GRU RNN, by $2 \times (n^2 + nm) = 2n(n + m)$.

$$i_t = \sigma(b_i) \tag{5.26}$$

$$o_t = \sigma(b_o) \tag{5.27}$$

The rational here is that the bias vectors get to be updated using the BPTT SGD using the co-state and indirectly the state in the process.

To reduce the parameters even further, one replaces the standard multiplications by point-wise (Hadamard) multiplications. In the case of the memory-cell units, the matrices U_* are reduced into (column) vectors of the same dimension as the memory-cell state units (i.e., n). We denote these corresponding vectors by u_* as delineated next.

5.3.4   Variant 4: The GRU_4 RNN

In this variant, each gate is computed using only the previous memory-cell state but with point-wise (Hadamard) multiplication with a compatible vector. Let u_* be an n-dimensional vector corresponding to the U_* matrix. Thus one reduces the total number of parameters, in comparison to the GRU RNN, by $2 \times (n^2 + nm) = 2n(n + m)$.

$$i_t = \sigma(u_i \odot c_{t-1}) \tag{5.28}$$

$$o_t = \sigma(u_o \odot c_{t-1}) \tag{5.29}$$

5.3.5   Variant 5: The GRU_5 RNN

In this variant, each gate is computed using the previous hidden state with point-wise (Hadamard) multiplication plus a bias as follows:

$$i_t = \sigma(u_i \odot c_{t-1} + b_i) \tag{5.30}$$

$$o_t = \sigma(u_o \odot c_{t-1} + b_o) \tag{5.31}$$

It is noted that there are more variants, similar to the LSTM RNN, which would attain more reductions in overall parameters for GRU RNN.

5.4   Sample Comparative Performance Evaluation

We have performed sample empirical experiments to illustrate the performance of the *standard* GRU RNN as well as to compare the performance among the standard GRU RNN with the first three variants. We employ the same datasets: first, sequences generated from the MNIST dataset and then on the IMDB movie review dataset. From here on, we will refer to the base GRU RNN model as GRU0 and the selected three variants as GRU1, GRU2, and GRU3, respectively. We shall leave the exploration of the performance of the other two variants as part of projects or exercises.

Table 5.1 Network characteristics

Network	MNIST pixel-wise	MNIST row-wise	IMDB
# Hidden units	100	100	128
Gate activation	Sigmoid	Sigmoid	Sigmoid
Activation	ReLU/tanh	ReLU/tanh	ReLU/tanh
Cost	Categorical cross-entropy	Categorical cross-entropy	Binary cross-entropy
# Epochs	100	100	100
Optimizer	RMSProp	RMSProp	RMSProp
Dropout	20%	20%	20%
Batch size	32	32	32

The network architecture adopted in this example case study consists of a single layer of one of the GRU RNN variants with an activation function g set as *ReLU* or *tanh*. For the MNIST dataset, we generate the pixel-wise and the row-wise sequences as in Chollet (b,c). The network architectures have been generated using the Keras library. We modified the Keras version of the GRU class to create classes for GRU1, GRU2, and GRU3. We trained and tested our networks using the *tanh*, and also separately using the *ReLU*, activation function on all these classes. The layer of units is followed by a 10-dimensional softmax layer in the case of the MNIST dataset or a traditional fully connected layer with scalar logistic activation in the case of the IMDB dataset to predict the probability of the output class. The Root Mean Square Propagation (RMSprop) is used as the choice of the optimizer to adapt each of the parameters. To speed up training, we also use the **exponential-loss learning rate**, see Chap. 2, which decays exponentially as the loss function decays. Let the loss function be L, then the (dynamic) learning rate in each epoch is given as

$$eta(n) = eta \times e^{\gamma L(n-1)} \tag{5.32}$$

where γ was set to 1, and *eta* represents the base constant learning rate, n is the current epoch number, $L(n-1)$ is the loss computed in the previous epoch, and $eta(n)$ is the current epoch (dynamic) learning rate. We trained the networks for 100 epochs in each case. The details of our networks are delineated in Table 5.1.

5.4.1 Application to MNIST Dataset (Pixel-Wise)

The classical MNIST dataset (LeCun et al., 2010) consists of 60,000 training images and 10,000 test images, each of size 28×28 of handwritten digits.

We describe here experiments for the evaluation of the *standard* GRU RNN vis-a-vis the first three variants on the MNIST dataset generated sequences. This is a typical approach as followed for the LSTM RNN evaluations in the previous chapter

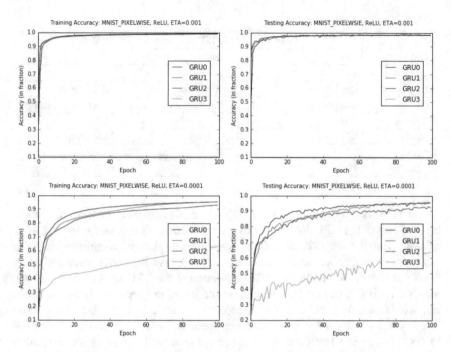

Fig. 5.1 Training and Testing accuracy on MNIST dataset with PIXEL-WISE sequences using ReLU activation function and base learning rates of $eta = 1e - 3$ and $eta = 1e - 4$

and moreover easily accessible via the computational frameworks and platforms. In one case, we generate the input sequence from each image *pixel-wise*, i.e., one pixel at a time. Each pixel-wise sequence generated from each image becomes 1-d signal sequence of length $28 \times 28 = 784$.

In the second case, we generate the input sequence from each image *row-wise*, i.e., one row at a time. Thus, for each image, the *row-wise* input produces a 28-d sequence of length 28.

For the pixel-wise case, we perform different experiments by varying the constant base learning rate *eta*. The summary results of the experiments are depicted in Figs. 5.1 and 5.2, with comparative summary in Tables 5.2 and 5.3, respectively.

From Tables 5.2, 5.3 and correspondingly Figs. 5.1 and 5.2, it can be observed that GRU1 and GRU2 perform comparably as well as GRU0 on MNIST pixel-wise generated sequence inputs. With further η tuning, the variants could improve their performance further. We are using these example experiments to demonstrate the capacity and the possibility, however. GRU3 did not perform as well for these two experiments using the specified (constant base) of the dynamic learning rates. In Table 5.2, in the case with *ReLU* activation and base learning rate of 0.001, GRU3 does not improve over the epochs. In contrast, in Table 5.3, the corresponding case with *tanh* activation function exhibits improvement. Figure 5.1 shows that reducing the (constant base) learning rate to 0.0001 enabled GRU3 to increase its (test) accuracy performance to 63.6% after 100 epochs, and the accuracy curves still have

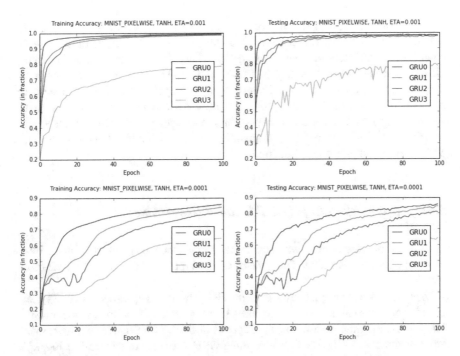

Fig. 5.2 Training and Testing accuracy on MNIST dataset with PIXEL-WISE sequence using TANH activation function and base learning rates of $eta = 1e - 3$ and $eta = 1e - 4$

Table 5.2 MNIST pixel-wise sequences, ReLU activation: percentage accuracy of different architectures using 2 constant base learning rates eta in 100 epochs

eta	10^{-3}		10^{-4}		
Variant	Train	Test	Train	Test	# Params
GRU0	99.2	98.6	95.3	95.3	30,600
GRU1	98.9	98.4	95.3	95.4	30,400
GRU2	98.9	98.1	92.9	92.2	30,200
GRU3	10.4	10.3	63.1	63.3	10,400

Table 5.3 MNIST pixel-wise sequences, TANH activation: percentage accuracy of different architectures using 2 constant base learning rates eta in 100 epochs

eta	10^{-3}		10^{-4}		
Variant	Train	Test	Train	Test	# Params
GRU0	99.2	98.3	85.8	85.7	30,600
GRU1	98.4	97.6	84.0	84.6	30,400
GRU2	98.8	98.1	80.6	80.7	30,200
GRU3	78.7	79.1	64.3	64.3	10,400

a positive slope indicating that it would increase further after additional epochs. We remark that in these experiments, GRU3 has about 33% of the number of (adaptively computed) parameters compared to GRU0. Thus, there exists a potential trade-off between the higher accuracy performance and the decrease in the number of parameters. In our example experiments, using 100 epochs, the GRU3 architecture never attains leveling off—indicating potential performance improvements with additional epochs. Further experiments using more epochs and/or more units would provide more insights into the comparative evaluation of this trade-off between performance and parameter-reduction.

5.4.2 Application to MNIST Dataset (Row-Wise)

In these set of experiments, we generate the input sequence from each image row-wise, i.e., one row at a time. Thus, the row-wise input produces a 28-d sequence of length 28.

While pixel-wise sequences represent relatively long sequences, row-wise generated sequences can test short sequences (here of length 28) with vector elements. The accuracy profile performance vs. epochs of the MNIST dataset with row-wise input of the base GRU RNN and the three GRU variants are depicted in Figs. 5.3

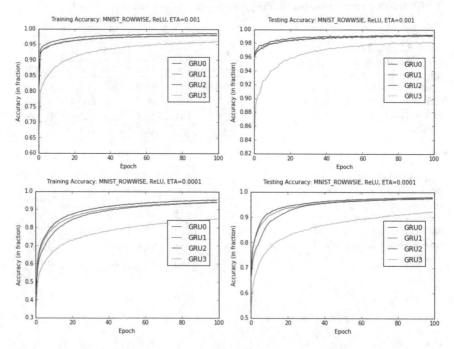

Fig. 5.3 Training and Testing accuracy on MNIST dataset with ROW-WISE sequences using ReLU activation function and base learning rates of $eta = 1e - 3$ and $eta = 1e - 4$

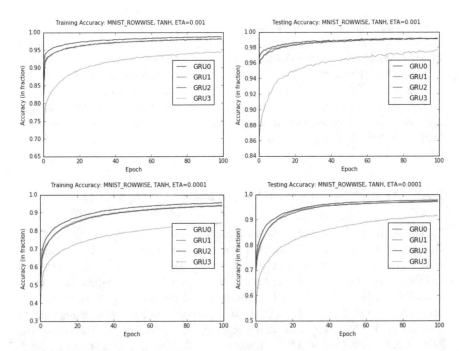

Fig. 5.4 Training and Testing accuracy on MNIST dataset with ROW-WISE sequences using TANH activation function and base learning rates of $eta = 1e - 3$ and $eta = 1e - 4$

Table 5.4 MNIST row-wise sequences, ReLU activation: percentage accuracy of different architectures using 2 base learning rates *eta* in 100 epochs

eta	10^{-3}		10^{-4}		
Variant	Train	Test	Train	Test	# Params
GRU0	98.5	99.2	95.3	97.8	38,700
GRU1	97.9	99.0	94.0	97.5	33,100
GRU2	98.0	99.1	94.0	97.4	32,900
GRU3	96.0	98.1	85.0	92.1	13,100

Table 5.5 MNIST row-wise sequences, TANH activation: percentage accuracy of different architectures using 2 base learning rates *eta* in 100 epochs

eta	10^{-3}		10^{-4}		
Variant	Train	Test	Train	Test	# Params
GRU0	98.7	99.1	95.4	97.7	38,700
GRU1	98.1	99.2	93.9	97.3	33,100
GRU2	98.1	99.1	93.6	97.0	32,900
GRU3	97.7	97.3	97.0	91.5	13,100

and 5.4, using two constant base values of the dynamic (exponential-loss) learning rate. The results of the accuracy performance are then summarized, respectively, in Tables 5.4 and 5.5.

From Tables 5.4, 5.5 and the corresponding Figs. 5.3 and 5.4, the base GRU RNN, i.e., GRU0, and all the three GRU variants, GRU1, GRU2, and GRU3, appear to exhibit comparable accuracy performance over two constant base learning rates. GRU3 exhibits low performance at the base learning rate of 1e-4 where, after 100 epochs, is still lagging. However, it appears that the profile has not yet leveled off and has a positive slope. More epochs are likely to increase performance to comparable levels with the other variants. It is noted here also that in this experiment, GRU3 can achieve comparable performance with roughly one third of the number of (adaptively computed) parameters. Computational expense savings may play a role in favoring one variant over the others in targeted applications and/or available resources at edge devices.

5.4.3   Application to the IMDB Dataset (Text Sequence)

The IMDB dataset is another classic dataset but which is naturally in sequence form. This dataset is composed of 25,000 training data and 25,000 test data consisting of movie reviews and their binary sentiment classification. Each review is represented by a maximum of 80 (most frequently occurring) words in a vocabulary (dictionary) of 20,000 words (Maas et al., 2011). We have used the dataset for training the base GRU RNN and all three GRU variants using the two constant base learning rates of 1e-3 and 1e-4 over 100 epochs. In the training, we employ 128-dimensional GRU RNN variants and have adopted a batch size of 32. Using a base learning rate of 1e-3, we have observed that performance fluctuates visibly, whereas it uniformly progresses over profile-curves when the base in the (exponential-loss) learning rate equals 1e-4, as shown in Fig. 5.5. Tables 5.6 and 5.7 summarize the results of accuracy performance in these experiments which show comparable performance among GRU0, GRU1, GRU2, and GRU3 with both *tanh* and *ReLU* activation functions. The number of parameters in each case is also listed in the two tables (Tables 5.6 and 5.7).

The IMDB data experiments provide the most striking results. It can be clearly seen that all the three GRU variants perform comparably to the base (standard) GRU RNN while using less number of parameters. The learning pace of GRU3 was also similar to those of the other variants at the constant base learning rate of 1e-4. From Tables 5.6 and 5.7, it is noted that more saving in computational load is achieved by all the variants of GRU RNN as the input is represented as a relatively large 128-dimensional vector.

5.5   Concluding Remarks

This chapter describes the original (standard) Gated Recurrent Unit (GRU) recurrent Neural Network (RNN) and contrasts it to the LSTM RNN with a common notation. GRU RNN uses one less gate by way of coupling two gates and adopting a "convex sum" in defining the gates in the "memory-cell." It uses the memory-cell as the feedback state instead of the activation function, and thus uses one (hyperbolic

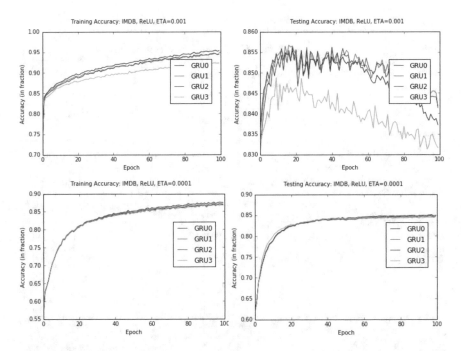

Fig. 5.5 Training and Testing accuracy on IMDB dataset using ReLU activation function and base learning rates of $eta = 1e - 3$ and $eta = 1e - 4$

Table 5.6 IMDB dataset, ReLU Activation: performance summary of different architectures using two base learning rates *eta* over 100 epochs

eta	10^{-3}		10^{-4}		
Variant	Train	Test	Train	Test	# Params
GRU0	95.3	83.7	87.4	84.8	98,688
GRU1	94.5	84.1	87.0	84.8	65,920
GRU2	94.5	84.2	86.9	84.6	65,664
GRU3	92.3	83.2	86.8	84.5	33,152

Table 5.7 IMDB dataset, TANH Activation: performance summary of different architectures using two base learning rates *eta* over 100 epochs

eta	10^{-3}		10^{-4}		
Variant	Train	Test	Train	Test	# Params
GRU0	96.9	83.4	88.3	85.2	98,688
GRU1	96.2	83.2	88.2	85.2	65,920
GRU2	95.8	83.6	88.0	85.1	65,664
GRU3	97	79	88	84.9	33,152

tangent) nonlinearity instead of two in the signal path. These two characteristics in the structure bring about savings in the computational load and consequently measurable speed up in training.

While retaining the GRU RNN structure, we also describe further computational savings by way of reducing parameters in the five *slim* GRU RNN variants. We then discuss sample experiments. The experiments on the original (standard) GRU RNN vis-a-vis the first three *slim* variants GRU1, GRU2, and GRU3 have demonstrated

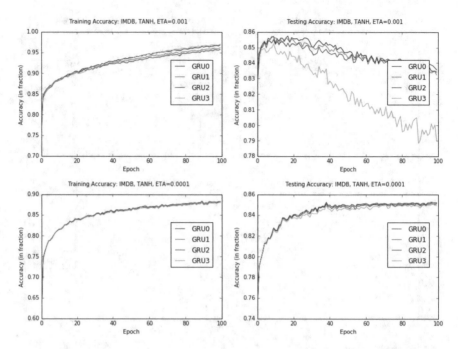

Fig. 5.6 Training and Testing accuracy on IMDB dataset using TANH activation function and base learning rates of $eta = 1e - 3$ and $eta = 1e - 4$

that their test accuracy performance is comparable on three example sequence lengths. Two sequences (artificially) generated from the *classical* MNIST dataset and one from the natural sequence of the (movie reviews) IMDB dataset. The main driving signal of the gates is the (recurrent) state as it contains essential information about the profile input sequence (Salem, 2016b). Moreover, the use of the stochastic gradient descent implicitly carries information about the network state (Salem, 2016a, 2016b). This may explain the relative success in using the bias alone in the gate signals as its adaptive update carries information about the state of the network (albeit implicitly). The *slim* GRU variants reduce redundancy, and thus their performance has been comparable to the original standard GRU RNN. While GRU1 and GRU2 have indistinguishable performance from the standard GRU RNN, GRU3—in these experiments—frequently lags in performance, especially for relatively long sequences and may require more execution time or larger base learning rate (eta) to achieve comparable performance.

By performing more experimental evaluations using constant or varying learning rates, and training for longer number of epochs, one can validate the performance on broader domains. We remark that the full *slim* GRU RNN variants should be comparatively evaluated on diverse datasets for broader empirical performance evidence. Such efforts are candidates for suitable projects to investigate the comparative performance of a host of reduced-structure and/or reduced-parameter variants on diverse datasets.

Chapter 6
Gated RNN: The Minimal Gated Unit (MGU) RNN

6.1 Introduction and Background

Various forms of recurrent neural networks (RNN) have been proposed since the Long Short-Term Memory (LSTM) form was published in 1997 with subsequent impressive results for numerous sequence-to-sequence applications (Chung et al., 2014b; Gers et al., 2002; Graves, 2012; Greff et al., 2017; Hochreiter & Schmidhuber, 1997; Johnson et al., 2016; Zaremba, 2015). Recently, simpler recurrent units with fewer gates and fewer parameters have shown comparable performances as the relatively more complicated architectures on example datasets (Akandeh & Salem, 2017a; Lu & Salem, 2017). Chung et al. (2014a) proposed a gated recurrent unit (GRU) in 2014 that uses two gates and can achieve comparable accuracies in example applications to the three-gated base LSTM RNN (Greff et al., 2017). Zhou et al. (2016) proposed a simpler minimal gated unit (MGU) based on a GRU that has only one single gate. In Zhou et al. (2016), the MGU-based RNN has shown similar accuracy as the GRU-based RNN on *example* datasets. The **MGU RNN**, with only a single gate, shows a simpler design and fewer parameters, and thus less training computational expense, potentially suitable for edge devices. The key question remains if these reduced structure forms can achieve comparable accuracy performance on broader datasets and applications.

The recent trend toward *simpler* recurrent units suggests a need for RNNs with smaller memory footprints and lower training computational load. The **MGU RNN** gated structure may not be further simplified, as it only has one dynamic gate Zhou et al. (2016) and Heck and Salem (2017). However, parameters used in the single gate could justifiably be eliminated to reduce **memory footprint** and **computational expense**.

In this chapter, we introduce the **(gated) MGU RNN** architecture which is induced from the base (i.e., standard) GRU RNN. While the MGU RNN is a simple induced form of the GRU RNN, it aspires to achieve comparable accuracy performance in *example* applications. We treat the MGU RNN as a separate form

F. M. Salem, *Recurrent Neural Networks*,
https://doi.org/10.1007/978-3-030-89929-5_6

(class) as we will further pursue reduction of parameters while retaining the form (class). Such choice enables the view of the **gated families**, i.e., LSTMs, GRUs, and MGUs, as gated architectures, each with its own generated set of *slim* variants focusing on reducing the internal parameters.

Following the same treatment of the LSTM and the GRU RNN, we also identify and introduce five *slim* **MGU RNNs**. Here, we call these *slim* variants, MGU1, MGU2, MGU3, MGU4, and MGU5, respectively. Then, we show the performance of the original MGU RNN as proposed by Zhou et al. (2016) in comparison to the first three *slim* variants. All three *slim* variants as well as the original MGU RNN have been comparatively evaluated on the standard sequences (artificially) generated from the MNIST dataset as well as the natural sequences of the **Reuters Newswire Topics** (RNT) dataset.

The remainder of the chapter is organized as follows: Sect. 6.2 presents a background on the **gated recurrent neural networks**, particularly employing the **minimal gated unit**, and introduces the *slim* variants. Section 6.3 specifies the network architectures and libraries used to evaluate the architectures. Section 6.4 comparatively summarizes the performance evaluation results of the standard MGU RNN and the *slim* variants using the two (MNIST and RNT) datasets. Finally, Sect. 6.5 summarizes concluding remarks.

In order to show the *transition* flow of (gated) architectures from the most general **LSTM RNN**, to the **GRU RNN**, then to the **MGU RNN**, to the **simple RNN**, we shall revisit these architectural descriptions for two reasons: (i) one can observe and relate the redesign *transitions* among the architectures, and (ii) render each chapter to become a self-contained study module, less dependent on other chapters. We shall describe the general background models next.

6.1.1   Simple RNN Architectures

For simple recurrent neural networks (Bengio et al., 1994; Goodfellow et al., 2016a; Pascanu et al., 2013), the recurrent vector state h_t is updated at each time step according to the following discrete dynamic model:

$$h_t = g(Uh_{t-1} + Ws_t + b) \tag{6.1}$$

where s_t is the external input vector, $g(\cdot)$ is usually chosen to be $tanh(\cdot)$, the hyperbolic tangent function, and the parameters are the matrices U, W, and the bias vector b, with appropriate sizes for compatibility. To be specific, we shall denote the dimensions of the input and recurrent state as m and n, respectively. Then U is an n-by-n matrix, W is an n-by-m matrix, and b is an n-by-1 vector.

6.1.2   LSTM RNN

Let us begin with the LSTM recurrent neural networks (RNNs) first as the more general (parent) architecture. LSTM RNNs have shown impressive results in several applications involving sequence-to-sequence (S-2-S) mappings, from speech recognition and translation to natural language processing, to biology and medical applications, Graves (2012), Greff et al. (2017), Johnson et al. (2016), Goodfellow et al. (2016a). They, however, possess more complex structure by introducing three-gated units and consequently increase the adaptive parameters by four-fold in comparison to the simple recurrent neural networks (sRNN) (Kent & Salem, 2019; Lu & Salem, 2017; Salem, 2018). It is appropriate to begin the gated RNN modeling transition from the LSTM RNN, which introduced the memory-cell unit structure with its associated gates as follows:

$$\tilde{c}_t = tanh(U h_{t-1} + W s_t + b) \tag{6.2}$$

$$c_t = f_t \odot c_{t-1} + i_t \odot \tilde{c}_t \tag{6.3}$$

$$h_t = o_t \odot g(c_t) \tag{6.4}$$

where $g(\cdot)$ is typically the hyperbolic tangent function $tanh(.)$, c_t is referred to as the (vector) memory-cell at time t, and $\tilde{c}_t$ is the candidate activation at t. The LSTM RNN in Eqs. (6.2)–(6.4) incorporates the sRNN model and the previous memory-cell value $c_{(t-1)}$ in an element-wise weighted sum using the *forget-gate* signal f_t and the *input-gating* signal i_t. Note that $\odot$ denotes element-wise (i.e., Hadamard) multiplication. Moreover, in the last equation in Eqs. (6.2)–(6.4), the memory-cell is passed through the activation function $tanh(\cdot)$ before multiplying it (element-wise) to the *output-gate* signal o_t to generate the hidden unit vector h_t.

Each of the three gate signals is obtained from a replica of the sRNN using the logistic activation, $\sigma(\cdot)$, to limit its gate signaling range to between 0 and 1. Specifically, the gate signals are expressed as:

$$i_t = \sigma(U_i h_{t-1} + W_i s_t + b_i) \tag{6.5}$$

$$f_t = \sigma(U_f h_{t-1} + W_f s_t + b_f) \tag{6.6}$$

$$o_t = \sigma(U_o h_{t-1} + W_o s_t + b_o) \tag{6.7}$$

where each dynamic (vector) gate signal has its own set of parameters. This constitutes an increase of (adaptively computed) parameters by four-folds in comparison to the sRNN. We relegate further details to Chap. 4 and the references therein.

6.1.3   GRU RNN

Recent research activities have sought to reduce LSTM structure complexity and/or reduce the number of parameters to minimize the required **memory print** and computational resources. The gated recurrent unit architecture (Chung et al., 2014b) is an example of such new structures with reduced number of gates.

We follow the original notation used when the GRU RNN was introduced. The gated recurrent units (Chung et al., 2014b) reduce the gated RNN to two gates: an *update-gate* z_t and a *reset-gate* r_t. The update-gate controls how much the unit updates its state, here called h_t as follows:

$$h_t = (1 - z_t) \odot h_{t-1} + z_t \odot \tilde{h}_t \qquad (6.8)$$

The reset-gate controls the amount of history used to update the candidate activation $\tilde{h}_t$, which would effectively become dependent on only the (external) input signal when the rest-gate is close to zero. Specifically,

$$\tilde{h}_t = tanh(U(r_t \odot h_{t-1}) + W s_t + b) \qquad (6.9)$$

The two gate equations, which are replicas of sRNN using the logistic function $\sigma(\cdot)$ for nonlinearity, have their own set of adaptive parameters as expressed in:

$$z_t = \sigma(U_z h_{t-1} + W_z s_t + b_z) \qquad (6.10)$$
$$r_t = \sigma(U_r h_{(t-1)} + W_r s_t + b_r) \qquad (6.11)$$

Thus, the GRU RNN increases the computationally adaptive parameters by three-folds in comparison to the sRNN.

6.1.4   Gated Recurrent Unit (GRU) RNNs vs. the LSTM RNNs

We can adopt the same notations as in the LSTM RNNs to express the equations for the GRU RNNs. One can easily assess the comparisons and see the structure reduction as applied to the LSTM RNN to obtain the GRU RNN. To that end, we use the following notation correspondence among the two gated RNNs:

$$i_t \longleftrightarrow z_t \qquad (6.12)$$
$$f_t \longleftrightarrow (1 - z_t) \qquad (6.13)$$
$$c_t \longleftrightarrow h_t \qquad (6.14)$$
$$\tilde{c}_t \longleftrightarrow \tilde{h}_t \qquad (6.15)$$

Thus, using the notation of LSTM RNNs, one can re-express the GRU RNNs as

$$\tilde{c}_t = g(U_c h_{t-1} + W_c s_t + b_c) \tag{6.16}$$

$$c_t = (1 - i_t) \odot c_{t-1} + i_t \odot \tilde{c}_t \tag{6.17}$$

$$h_{t-1} = o_t \odot c_{t-1} \tag{6.18}$$

where one observes that the gating in the memory-cell is using a single gate, i_t, and its (convex) complement $1 - i_t$. The new activation in Eq. (6.18) has dropped the nonlinearity g in comparison and uses the memory-cell state directly. Thus, the architecture path passes through only one g nonlinearity in each time cycle. Moreover, there is inconsistent time-delay in the expression in Eq. (6.18).

To make the GRU RNN a direct reduced form of the original LSTM RNN, one may revise the activation equation (6.18) to have all signals at the same instant of time as:

$$h_t = o_t \odot c_t$$

The gate signals i_t and o_t are similar to the one in the LSTM RNN architecture, with a re-focus on the memory-cell state feedback c_t (as opposed to the activation state feedback h_t), specifically,

$$i_t = \sigma(U_i c_{t-1} + W_i s_t + b_i) \tag{6.19}$$

$$o_t = \sigma(U_o c_{t-1} + W_o s_t + b_o) \tag{6.20}$$

6.2 The Standard MGU RNN

The **minimal gated unit RNN** proposed in Zhou et al. (2016) reduces the number of gates in a GRU RNN from two to one by basically using (or sharing) the *update or input* gate with the *reset or output* gate. This sharing results in one gate, which is labeled as the input gate i_t below, and is computed in the same way as above. Specifically,

$$i_t = \sigma(U_i c_{t-1} + W_i s_t + b_i) \tag{6.21}$$

As compared to the GRU RNN, the MGU RNN equations then become:

$$\tilde{c}_t = g(U_c(i_t \odot c_{t-1}) + W_c s_t + b_c) \tag{6.22}$$

$$c_t = (1 - i_t) \odot c_{t-1} + i_t \odot \tilde{c}_t \tag{6.23}$$

$$i_t = \sigma(U_i c_{t-1} + W_i s_t + b_i) \tag{6.24}$$

where g is typically the hyperbolic tangent function $tanh(\cdot)$. Thus, this constitutes an increase of (adaptive) parameters to only two-folds in comparison to the sRNN. The MGU RNN has approximately 33 % fewer (adaptive) parameters than the GRU RNN. If the level of accuracy performance in certain applications is retained, this parameter reduction may become attractive for pressed resources of edge devices.

6.3   Slim MGU RNN: Reductions within the Gate(s)

In this section, for completeness in reference to Chaps. 4 and 5, we explicitly identify the five *slim* variants similar to the ones we introduced for both the LSTM and GRU RNN.

The *slim* MGU variants considered here provide further simplicity by reducing the number of adaptive parameters (in the single gate eqn) and consequently further reduce the computational expense. The MGU variants introduced here are called simply MGU_1, MGU_2, MGU_3, MGU_4, and MGU_5. (Observe that the input signal s_t is eliminated from the gate equation in all the 5 variants. For justification for this choice, we refer the reader to remarks and points made in Chap. 4.)

6.3.1   Variant 1: MGU_1 RNN

The first variation on the MGU RNN architecture is to remove the input signal s_t from the gate signal equation, making the gate dependent only on the unit history and bias.

$$i_t = \sigma(U_i c_{t-1} + b_i) \tag{6.25}$$

This variation reduces the number of parameters by the size of the matrix W_i which is equal to n-by-m, or nm parameters in comparison to the original MGU RNN model.

6.3.2   Variant 2: MGU_2 RNN

The second variation is to remove the input signal s_t and the bias b_i from the gate equation, making the gate dependent only on the memory-cell, i.e., state, history.

$$i_t = (U_i c_{(t-1)}) \tag{6.26}$$

This variation further reduces the parameters by the n elements of b_i than the MGU_1 variant. In total, the reduction equals $n(m+1)$ in comparison to the original MGU RNN.

6.3.3   Variant 3: MGU_3 RNN

The third variation is to remove the input signal s_t and the cell-memory state history c_{t-1}, leaving just the bias term.

$$i_t = \sigma(b_i) \tag{6.27}$$

This variation reduces the parameters by $n(n + m)$ in comparison to the original MGU RNN. In the experiments shared below, this variation has about 50% of the parameters compared to the original MGU RNN. Thus, the memory footprint would be lower, and the training and execution would be much faster.

To reduce the parameters even further, one replaces the standard matrix multiplication by point-wise (Hadamard) multiplications. In the case of the memory-cell units, the matrix U_i is reduced into a (column) vector of the same dimension as the hidden units (i.e., n). We denote this corresponding vector by u_i as delineated next.

6.3.4   Variant 4: MGU_4 RNN

In this variant, the gate is computed using only the previous memory-cell state but with point-wise multiplication. Thus one reduces the total number of parameters, in comparison to the original MGU RNN, by $(n^2 + nm) = n(n + m)$.

$$i_t = \sigma(u_i \odot c_{t-1}) \tag{6.28}$$

6.3.5   Variant 5: MGU_5 RNN

In this variant, the gate is computed using the previous hidden state with point-wise multiplication plus a bias. Thus one reduces the total number of parameters, in comparison to the original MGU RNN, by $(nm + n^2 - n) = n(n + m - 1)$.

$$i_t = \sigma(u_i \odot c_{t-1} + b_i) \tag{6.29}$$

We remark that we followed the same pattern and naming style for all the gated RNNs. These *slim* gated RNNs focus on the gate signals only. They apply the same process in the parameter reductions within a gate.

6.4   Sample Comparative Performance Evaluation

6.4.1   Sample Comparative MGU RNN Performance

Zhou et al. (2016) performed computational experiments on MGU RNN on sequences generated from the MNIST dataset and reported comparable accuracies to GRU and LSTM RNNs. The generated (artificial) sequences were formed, as before, by converting each 28x28 MNIST image, row-wise, to a 28-element vector of 28-length sequences. Analogously, they rolled out the 28x28 image row-wise into a single vector of 784-length sequence. The generated sequences of length 28 or 784 were employed to train 100 hidden MGU RNN over thousands of epochs. Their results for 28-length sequences showed a testing accuracy of 88% for MGU after 16,000 epochs using a batch size of 100. For 784-length sequences, they reported a testing accuracy of 84.25% after 16,000 epochs. These test results were comparable to the base GRU RNN under equivalent training and testing conditions.

In this section, we shall conduct experiments on the same two datasets to evaluate the performance of the MGU RNN in comparison to three of its *slim* MGU variants, namely, MGU_1, MGU_2, and MGU_3 (Heck and Salem, 2017). We shall leave the other two variants for projects and further case studies by the motivated reader. We shall proceed next by describing the neural network architecture used in this case study.

6.4.2   The Network Architecture

The neural network architecture chosen for the experiments using the two datasets, the MNIST and the RNT datasets, were created employing the Keras deep learning library/framework. Since Keras has a GRU layer class, this class was modified to create classes for the MGU and the MGU_1, MGU_2, and MGU_3 variants. All these 4 classes used the hyperbolic tangent function for the candidate activation, and the logistic sigmoid function for the gate activation.

For the MNIST dataset, we used a batch size of 100 and the RMSProp optimizer. A single layer of hidden units was used with 100 units for the 784-length sequences and 50 units for the 28-length sequences. Although (Zhou et al., 2016) used 100 units for both lengths of sequences, for our experiments we decreased the number of units for the shorter sequences as that was sufficient to achieve reasonable comparative (test) accuracy performance and also to decrease training time (Heck & Salem, 2017). The output layer was a fully connected feedforward network (fc) layer of 10 units in both cases. Table 6.1 summarizes the number of (adaptive) parameters used in the MGU, MGU_1, MGU_2, and MGU_3 for the case studies for the sequences generated from the MNIST dataset. In the following tables and the figures, we refer to the names as MGU0, MGU1, MGU2, and MGU3, respectively.

Table 6.1 Number of Parameters in the MNIST row-wise vs. the MNIST pixel-wise Network Architectures

	Row-wise	Pixel-wise
Unit/input/length	50/28/28	100/1/784
Variant	Parameters	
MGU0	7900	20,400
MGU1	6500	20,300
MGU2	6450	20,200
MGU3	4000	10,300

Table 6.2 Number of parameters in the RNT network architecture

	Unit/input/length
Variant	250/1/500
MGU0	126,000
MGU1	125,750
MGU2	125,500
MGU3	63,250

For the MNIST dataset, the 28-length sequence architectures were run for 50 epochs, whereas the 784-length sequence architectures were run for 25 epochs to decrease training time for the longer sequences. Both networks were trained on multiple learning rates for the RMSProp optimizer: 10^{-3}, 10^{-4}, and 10^{-5}.

The RNT dataset was evaluated using a sequence length of 500, a single hidden layer of 250 units, and a batch size of 64. The output layer contained 46 fully connected units for the news categories. Other combinations of sequence length and hidden units were tested, and the best results were with a ratio of 2-to-1. A sequence length of 500 with 250 hidden units was chosen (Heck & Salem, 2017). Instead of RMSProp, the Adam optimizer (Chollet, a) was used as it provided slightly better performance results. The learning rate was the default 10^{-3} as used in Zhou et al. (2016). The network was trained across 30 epochs, which was sufficient to infer the comparative performance in (testing) accuracy. Table 6.2 summarizes the (adaptive) parameters used in the MGU and its *slim* variants when using 250 units with sequence length of 500.

We now summarize the experimental results of the original MGU RNN and its *slim* variants on the MNIST and RNT datasets. The original MGU RNN performance results serve as a baseline for comparison to the three *slim* MGU variants.

6.4.3   Comparative Performance on the MNIST Dataset

The MNIST database contains 28×28-pixel grayscale images of handwritten digits between zero and nine (Chollet, a; LeCun et al., 2010). These images are separated into a training set of 60,000 images and a test set of 10,000 images. Following procedures in Zhou et al. (2016), Chollet (b,c) and the references therein, the dataset is evaluated by treating each image as a sequence of 28 elements, each of size (dimension) 28, and alternatively, as a sequence 784 elements, each of size (dimension) one. Using these two varied sequence representations allows for more comparison of the results of the MGU RNN and the three *slim* MGU variants. The MNIST database was retrieved from the Keras deep learning platform library (Chollet, a).

(1) The 784-Length Generated Sequence
The best performance on the 784-length MNIST data resulted from a learning rate of 10^{-3}. Initial performance with that learning rate was inconsistent with significant spikes in the accuracies until the later epochs, as shown in Fig. 6.1. For most of the epochs, MGU2 had the best accuracy, and it achieved slightly better accuracy than MGU after only 25 epochs. The consistent result on this dataset was the relatively poor performance of MGU3 with a learning rate of 10^{-3}. For a relatively low epoch of 25 and with a learning rate of 10^{-4} and 10^{-5}, MGU3 achieved accuracies similar to the other variants, as shown in Table 6.3. This result suggests a need to further verify the performance of the MGU3 variant on more case studies and longer epoch runs beyond the scope and the sample datasets in this study. It is noted that these studies using the MNIST dataset are referred to as "toy" problems, and their values

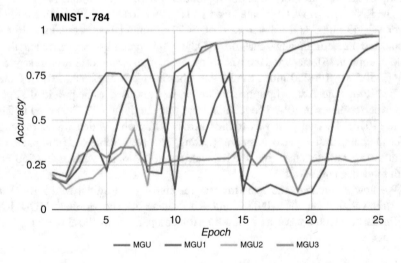

Fig. 6.1 Results of MNIST pixel-wise 784-length sequence with a learning rate of $eta = 1e - 3$

Table 6.3 Accuracy for MNIST pixel-wise after 25 epochs	Learning rate	10^{-3}	10^{-4}	10^{-5}
	Variant	Accuracy		
	MGU0	96.8	34.7	20.2
	MGU1	92.8	40.8	21.8
	MGU2	97.1	42.2	21.1
	MGU3	29	40.8	21.1

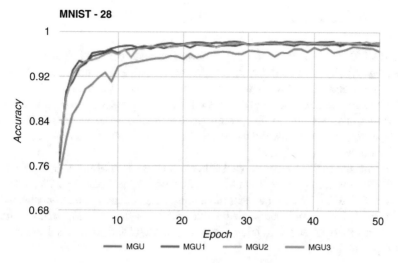

Fig. 6.2 results of MNIST row-wise 28-length sequence with a learning rate $\eta = 1e^{-3}$

Table 6.4 Accuracy for MNIST row-wise after 50 epochs

Learning rate	10^{-3}	10^{-4}	10^{-5}
Variant	Accuracy		
MGU0	97.6	95.6	69.5
MGU1	98.1	95.3	71.3
MGU2	98.2	94.2	71.2
MGU3	96.6	91.6	65.5

are mainly to establish a baseline comparison among the different variant gated RNN architectures.

(2) The 28-Length Generated Sequence

The accuracy performance on the 28-length sequence MNIST data was relatively high after just 50 epochs. Figure 6.2 shows that the accuracy is above 90% after just several epochs for all variants; however, MGU3 still performs relatively lower. The highest performance resulted from a learning rate of $1e^{-3}$, although a lower rate of $1e^{-4}$ was only slightly worse. Summary performance of the MGU RNN and its three *slim* variants are tabulated for a grid of three learning rate values in Table 6.4.

Overall, for our hyper-parameter choices, MGU, MGU1, and MGU2 have competitive performances, whereas MGU3 would require either different hyper-parameter settings and/or longer epochs.

For two of the learning rates tested, including the best performance learning rate, MGU1 and MGU2 outperformed MGU by at least 0.5%. MGU3 performed relatively well even though it only contains the bias term in the gate equation. These initial learning rate grid choices indicate that increasing the learning rate higher than $1e^{-3}$ may improve performance of all variants.

6.4.4   Reuters Newswire Topics Dataset

The RNT database is a set of 11,228 newswire texts collected from the Reuters news agency. These newswires belong to 46 classes based on their topics. The training set includes 8982 newswires, and the testing set includes 2246. Each newswire in the database is preprocessed into a sequence of word indexes with an index corresponding to the overall frequency of a word in the database. For the following experiments, the dataset was retrieved from the Keras deep learning library framework (Chollet, a).

   As with the MNIST dataset, MGU2 performed the best of the MGU variants on the RNT dataset, improving upon the accuracy of original MGU by 22%, as shown in Table 6.5. MGU2 also featured a more consistent accuracy across epochs, unlike the other variants which had some notable spikes, as shown in Fig. 6.3. It is noted that the average per-epoch training time for each MGU variant would decrease with fewer parameters.

Table 6.5 Accuracy for RNT Net after 30 epochs

Learning rate	10^{-3}
Variant	Accuracy
MGU0	46
MGU1	49.2
MGU2	56.2
MGU3	39.7

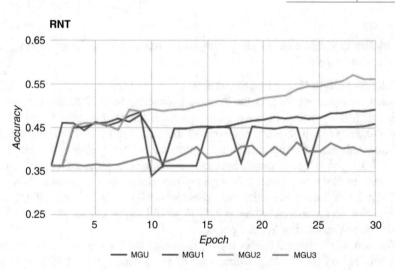

Fig. 6.3 Results of the RNT Network with a learning rate of $eta = 1e^{-3}$

6.4.5 Summary Discussion

The MGU RNN (MGU0) and the three *slim* MGU variants (MGU1, MGU2, and MGU3) were trained and evaluated on two datasets with different types of generated sequence data (image and text) and different lengths (28, 500, and 784). Compared to MGU0, MGU2 provided better accuracy for both datasets. Since MGU2 does not include the input signal or bias in the (single) gate equation, it achieved this performance with fewer parameters. MGU2 also displayed better accuracy than the other two *slim* MGU variants. However, MGU3, which has 50% fewer parameters than MGU0, achieved similar, albeit worse, accuracy performance to MGU0 in two of the tests. This result suggests that a network with the MGU3 gate structure could work reasonably well in certain applications for which a smaller footprint is more critical than achieving the best accuracy.

6.5 Concluding Remarks

We described the original MGU RNN and how it is formed by simply sharing the update and reset gates of the GRU RNN. As was done in the LSTM and GRU RNNs, we then introduced five *slim* MGU RNN variants that incrementally reduce parameters within the (single) gate. These five variants were defined by reducing the number of parameters in the update-gate equation in the original MGU. We simply call these MGU RNN variants MGU1, MGU2, MGU3, MGU4, and MGU5. We then describe example experiments using MGU1, MGU2, and MGU3 and show their comparable (evaluation/testing) performance to the original MGU RNN (MGU0) on two popular datasets. The *slim* variant MGU2 achieved higher accuracy than the original MGU with fewer parameters and consequently lower training load expense. Since MGU2 can achieve better performance than MGU, which (Zhou et al., 2016) reported to achieve comparable performance to GRU, MGU2 may be used in some applications in place of GRU and MGU to achieve similar accuracy and at lower computational expense (Heck & Salem, 2017). Even MGU3 may be used in some cases if fewer parameters and faster training were more important than the highest accuracy performance.

These gated RNNs and their introduced *slim* variants may be used in projects or applications where varied level of computational resources or constraints are in place. Moreover, projects can explore more diverse datasets to gain a better understanding of how the gated RNNs and their *slim* variants perform compared to MGU RNNs across a diverse range of sequence domain applications. The reader should consult the dedicated **book repository** for experimental runs and evolving student projects using these varied **gated and un-gated recurrent neural networks**.

References

Ahmad, M., & Salem, F. M. (1992). Dynamic learning using exponential energy functions. In *Proceedings of the IEEE International Joint Conference on Neural Networks* pp. (II–121–II–126).

Akandeh, A., & Salem, F. M. (2017a). Simplified long short-term memory recurrent neural networks: part I. arXiv:1707.04619.

Akandeh, A., & Salem, F. M. (2017b). Simplified long short-term memory recurrent neural networks: Part II. arXiv:1707.04623.

Akandeh, A., & Salem, F. M. (2017c). Simplified long short-term memory recurrent neural networks: Part III. arXiv:1707.04626.

Albataineh, Z., & Salem, F. M. (2016). Adaptive blind CDMA receivers based on ICA filtered structures. *Journal of Circuits Systems & Signal Process* 1411–1446. https://doi.org/doi:10.1007/s00034-016-0459-4

Antsaklis, P., & Michel, A. (2006). *Linear systems*. Birkhauser.

Bengio, Y., LeCun, Y., Hinton, G. (2015). Deep learning. *Nature, 521*, 436–444.

Bengio, Y., Simard, P., & Frasconi, P. (1994). Learning long-term dependencies with gradient descent is difficult. *IEEE Transactions on Neural Networks, 5*(2), 157–166.

Bertsekas, D. (2019). *Reinforcement learning and optimal control*. Athena Scientific.

Boulanger-Lewandowski, N., Bengio, Y., & Vincent, P. (2012). Modeling temporal dependencies in high-dimensional sequences: Application to polyphonic music generation and transcription. arXiv preprint arXiv:1206.6392.

Boyd, S., & Vandenberghe, L. (2004). *Convex optimization*. Cambridge University Press. https://web.stanford.edu/~boyd/cvxbook/

Boyd, S., & Vandenberghe, L. (2018). *Introduction to applied linear algebra—vectors, matrices, and least squares*. Cambridge University Press. https://web.stanford.edu/~boyd/vmls/

Bryson, Jr. A. E., & Y-C. Ho. (1975). *Applied optimal control: Optimization, estimation and control*. CRC Press.

Chollet, F. Keras. https://keras.io

Chollet, F. Keras-codes2. https://github.com/keras-team/keras

Chollet, F. Keras-codes. https://github.com/keras-team/keras

Chung, J., Gulcehre, C., Cho, K., & Bengio, Y. (2014a). Empirical evaluation of gated recurrent neural networks on sequence modeling. https://arxiv.org/abs/1412.3555

Chung, J., Gulcehre, C., Cho, K., & Bengio, Y. (2014b). Empirical evaluation of gated recurrent neural networks on sequence modeling. arXiv preprint arXiv:1412.3555

Dey, R., & Salem, F. M. (2017a). Gate-variants of gated recurrent unit (GRU) neural networks. In *2017 IEEE 60th International Midwest Symposium on Circuits and Systems (MWSCAS)* (p. 1597).

Dey, R., & Salem, F. M. (2017b). Gate-variants of gated recurrent unit (GRU) neural networks. arXiv preprint arXiv:1701.05923.

Fukushima, K. (1975). Cognitron: A self-organizing multilayered neural network. *Biological Cybernetics, 20,* 121–136.

Fukushima, K. (1980). Neocognitron: A self-organizing neural network model for a mechanism of pattern recognition unaffected by shift in position. *Biological Cybernetics, 36,* 193–202.

LeCun, Y., Bottou, L., Bengio, Y., & Haffner, P. (1998). Gradient-based learning applied to document recognition. *Proceedings of the IEEE, 86,* 2278–2324.

Gers, F. A., Schmidhuber, J., & Cummins, F. (2000). Learning to forget: Continual prediction with LSTM. *Neural Computation, 12*(10), 2451–2471.

Gers, F. A., Schraudolph, N. N., & Schmidhuber, J. (2002). Learning precise timing with LSTM recurrent networks. *Journal of Machine Learning Research, 3,* 115–143.

Goodfellow, I., Bengio, Y., & Courville, A. (2016a). *Deep learning.* MIT Press. http://www.deeplearningbook.org

Goodfellow, I., Bengio, Y., & Courville, A. (2016b). *Deep learning.* MIT Press.

Graves, A. (2012). Supervised sequence labelling with recurrent neural networks. In *Studies in Computational Intelligence.* Springer.

Greff, K., Srivastava, R. K., Koutník, J., Steunebrink, B. R., & Schmidhuber, J. (2017). LSTM: A search space odyssey. *IEEE Transactions on Neural Networks and Learning Systems, 28*(10), 2222–2232.

Haykin, S. (2009). *Neural networks and learning machines.* Prentics Hall.

Heck, J., & Salem, F. M. (2017). Simplified minimal gated unit variations for recurrent neural networks. In *2017 IEEE 60th International Midwest Symposium on Circuits and Systems (MWSCAS)* (p. 1593).

Hochreiter, S., & Schmidhuber, J. (1997). Long short-term memory. *Neural Computation, 9*(8), 1735–1780.

Jaitly, N., Le, Q. V., & Hinton, G. E. (2015). A simple way to initialize recurrent networks of rectified linear units. https://arxiv.org/abs/1504.00941

Jianxin, C.-L., Zhang, G.-B., Zhou, Wu, J., & Zhou, Z.-H. (2016). Minimal gated unit for recurrent neural networks. https://arxiv.org/abs/1603.09420.

Johnson, M., Schuster, M., Le, Q. V., Krikun, M., Wu, Y., Chen, Z., Thorat, N., Viégas, F. B., Wattenberg, M., Corrado, G., M. Hughes, & Dean, J. (2016). Google's multilingual neural machine translation system: Enabling zero-shot translation. http://arxiv.org/abs/1611.04558.

Kandel, E., Koester, J. D., Mack, S. H., & Sieglbaum, S.. (2021). *Principles of neural science* (6th ed.). McGraw Hill.

Kent, D., & Salem, F.M. (2019). Performance of three slim variants of the long short-term memory (LSTM) layer. In *2019 IEEE 62nd International Midwest Symposium on Circuits and Systems* (pp. 307–310). IEEE.

Kingma, D., & Ba, J. (2015). Adam: A method for stochastic optimization. In *3rd International Conference for Learning Representations.* https://arxiv.org/abs/1412.6980

Le, Q. V., Jaitly, N., & Hinton, G. E. (2015). A simple way to initialize recurrent networks of rectified linear units. arXiv preprint arXiv:1504.00941

LeCun, Y., Cortes, C., & Burges, C.J. C. (2010). Mnist handwritten digit database. In *AT&T labs.* http://yann.lecun.com/exdb/mnist

Lu, Y., & Salem, F. M. (2017). Simplified gating in long short-term memory (LSTM) recurrent neural networks. In *2017 IEEE 60th International Midwest Symposium on Circuits and Systems (MWSCAS)* (p. 1601).

Maas, A. L., Daly, R. E., Pham, P. T., Huang, D., Ng, A. Y., & Potts, C. (2011). Learning word vectors for sentiment analysis. In *Proceedings of the 49th Annual Meeting of the Association for Computational Linguistics: Human Language Technologies-Volume 1* (pp. 142–150). Association for Computational Linguistics.

Mikolov, T., Joulin, A., Chopra, S., Mathieu, M., & Ranzato, M. A. (2014). Learning longer memory in recurrent neural networks. arXiv preprint arXiv:1412.7753.

Pascanu, R., Mikolov, T., & Bengio, Y. (2013). On the difficulty of training recurrent neural networks. *ICML (3)*, *28*, 1310–1318.

Riley, K., Hobson, M., & Bence, S. (2006). *Mathematical methods for physics and engineering: A comprehensive guide* (3rd ed.). Cambridge University Press.

Salem, F. M. (2016a). A basic recurrent neural network model. arXiv preprint arXiv:1612.09022.

Salem, F. M. (2016b). Reduced parameterization in gated recurrent neural networks. Technical Report 11-2016, MSU.

Salem, F. M. (2018). Slim LSTMs. https://arxiv.org/abs/1812.11391

Tieleman, T., & Hinton, G. (2012). Lecture 6.5-rmsprop: Divide the gradient by a running average of its recent magnitude. In *COURSERA: Neural networks for machine learning*. https://www.cs.toronto.edu/~tijmen/csc321/slides/lecture_slides_lec6.pdf

Vrabie, D., Lewis, F., & Syrmos, V. (2012). *Optimal control*. Wiley.

Waheed, K., & Salem, F. M. (2003). Blind source recovery: A framework in the state space. *Journal of Machine Learning Research, 4*, 1411–1446.

Zaremba, W. (2015). An empirical exploration of recurrent network architectures. In *An empirical exploration of recurrent network architectures*.

Zaremba, W., Jozefowicz, R., & Sutskever, I. (2014). An empirical exploration of recurrent network architectures. https://arxiv.org/abs/1412.3555

Zaremba, W., Sutskever, I., & Vinyals, O. (2014). Recurrent neural network regularization. arXiv preprint arXiv:1409.2329.

Zhou, G.-B., Wu, J., Zhang, C.-L., & Zhou, Z.-H. (2016). Minimal gated unit for recurrent neural networks. *International Journal of Automation and Computing, 13*(3), 226–234

Index

A

Activation 43
Activation state 86
Adam 26
Adaptive Least-Mean-Square 38
Adaptive non-convex optimization 43
Amachrine cells (layer) 5
Approximate dynamic programming 23
Artificial intelligence viii, xvii, 7

B

Backpropagation through time (BPTT) 43, 45, 47, 61
Basic recurrent neural network (bRNN) 43, 49, 71
Binary cross-entropy 29, 80
Bipolar cells layer 5
Book repository 113
Bounded-input-bounded-state (BIBS) 49, 60

C

Calculus 55
Calculus of Variations 47, 50, 52
Clipping 48
Computational expense 101
Computational framework 78
Conditional independence xv
Convolutional neural networks 15
Co-state 50, 52–54
Coupling 85, 88
Covariance xv
Cross entropy 28, 56

D

Derivative xv
Determinant xiv
Difference equation 62
Difference Liapunov function 63
Differential equation 7
Directional derivative 24
Direct-sum 86
Discrete dynamic equations 87
Dropout layer 8
Dynamic learning rate 79

E

Effective learning rate *see* Dynamic learning rate
Element-wise product *see* Hadamard product
Ellipsoid 63
Entropy 49, 56
Epochs 39, 57
Exploding gradient 47, 60
Exponential-loss learning rate 26, 93

F

Finite-horizon 12, 43

G

Ganglion cells layer 5
Gated families *see* Gated RNN
Gated Recurrent Neural Networks 85
Gated Recurrent Unit (GRU) 85
Gated Recurrent Unit (GRU) RNNs 85, 87

Gated RNN 54, 71, 75, 85
Gated RNN variants 71
Gate-networks *see* Gtes 85
Gate-variants 86
Gating signaling *see* Gtes 85
Gating signals 73
Generalized backpropagation through time 61
Graph xiv

H
Hadamard multiplication 55, 57
Hadamard product xiv
Hebbian learning 22
Hessian matrix xv
Horizontal cells (layer) 5
Hyperbolic tangent 7
Hyper-parameter 45

I
Identity matrix 63
IMDB dataset 79, 86
IMDB movie review dataset *see* IDB dataset 86
Independence xv
Independent component analysis (ICA) 22
Inference 54, 59
Inference modes 75
Integral xv
Internal memory state 72
Iteration 57

J
Jacobian matrix xv

K
Keras 80
Kernel size 8, 15
Kullback–Leibler divervence xv, 28

L
Lagrange multiplier 47, 49, 52
Laplacian density function 56
Lateral connection 5
Learning 85
Learning-rate 45
Least Mean Square (LMS) 27
Liapunov function 62
Liapunov theory 63
Logistic function 7

Long-term dependencies 72
Long Short-Term Memory (LSTM) 24, 71, 85
Loss function 21, 24
LSTM RNNs 85

M
Matrix xiii, xiv
Memory cell 71, 72
Memory-cell state 86
Memory footprint 101
Memory print 104
MGU RNN 101, 105
Minimal gated unit (MGU) *see* MGU RNN
(gated) MGU RNN *see* MGU RNN
Mini-batch 39, 54, 57, 58
MNIST dataset 78
(Hadamard) multiplication 72

N
Negative definite 63, 64
Neural networks and deep learning 24, 27
Non-convex constrained optimization 49
Nonlinear dynamical systems 90
Norm xv

O
Online learning 58
Optic nerve 5
Optimizer Adam 80

P
Pixel-wise 86
Principal component analysis (PCA) 22

R
Receptive field 15
Rectified linear unit 7, 43
Recurrent neural networks (RNNs) 71, 85
Regression 49
Regularization 56
Reinforcement learning (RL) 23
Reset gate 88
Retina 5
Reuters Newswire Topics (RNT) 101, 102
RMSprop 26
Row-wise 86

S
Scalar xiii, xiv
Sensitivity 46, 47, 53

Sequence-to-sequence 60
Set xiv
Shannon entropy xv
Sigmoid xv, *see* Logistic function
Signal path 86
Simple RNN 43, 71
Simple RNN (sRNN) *see* sRNN
slim GRU RNNs 85, 86
slim MGU RNNs 102
Softmax 27
Softplus xv
Sparsity 27, 49, 56
Spiking neuron 7
Split boundary conditions 53, 64
Square matrix 62
sRNN 43, 71
Standard LSTM RNN 71
Statistical independence 27
Stochastic gradient descent 22, 43
Stopping criterion 54
Sup-Gaussian 56
Surface direction 25
Symmetric matrix 63

Symmetric positive definite matrix 62
System and signal processing 43

T
Tensor xiii, xiv
The stochastic gradient descent (SGD) 24
Transfer function 7
Transpose xiv

U
Unsupervised loss function 56
Update gate 88

V
Vanishing and exploding 60
Vanishing gradient 47, 59
Variance xv
Vector xiii, xiv
Vector-calculus 55
Visual cortex 5

Printed in the United States
by Baker & Taylor Publisher Services